Praise for *Love and Math*

"[Frenkel's] winsome new memoir... is three things: a Platonic love letter to mathematics; an attempt to give the layman some idea of its most magnificent drama-in-progress; and an autobiographical account, by turns inspiring and droll, of how the author himself came to be a leading player in that drama.

The conviction that mathematics has a reality that transcends the human mind is not uncommon among its practitioners, especially great ones like Frenkel and Langlands, Sir Roger Penrose and Kurt Gödel. It derives from the way that strange patterns and correspondences unexpectedly emerge, hinting at something hidden and mysterious."

—Jim Holt, *The New York Review of Books*

"With every page, I found my mind's eye conjuring up a fictional image of the book's author, writing by candlelight in the depths of the Siberian winter like Omar Sharif's Doctor Zhivago in the David Lean movie adaptation of Pasternak's famous novel. *Love and Math* is Edward Frenkel's *Lara poems*... As is true for all the great Russian novels, you will find in Frenkel's tale that one person's individual story of love and overcoming adversity provides both a penetrating lens on society and a revealing mirror into the human mind."

—Keith Devlin, *Huffington Post*

"Frenkel writes that math 'directs the flow of the universe.' It's as elegant as music and as much a part of our intellectual heritage as literature. He strives to awaken our wonder by taking us on [a] tour of his research, in which he reveals a 'hidden' world few of us encountered in school... Frenkel aims to make it understandable, even beautiful."

—*The New York Times Book Review*

"The words love and math aren't usually uttered in the same breath. But mathematician Edward Frenkel is on a mission to change that... [in his] book, 'Love and Math' [in which] the tenured professor at the University of California at Berkeley argues that the boring way that math is traditionally taught in schools has led to a widespread ignorance that may have even been responsible for the recession... [the] book tells his personal story and goes on to describe his research in the Langlands program, as well as recent mathematical discoveries that aren't regularly taught in classrooms."

—*The Wall Street Journal*

"The story of [Frenkel's] professional triumph against heavy odds is deeply satisfying... But his true answer to the bigotry he encountered in his youth lies in his passion for mathematics – the 'love' of the book's title... Believing that mathematics is a common human possession, he explains each concept in nontechnical terms, relying heavily on analogies from daily life... lay readers... will gain an understanding of what modern mathematics is about – its ambition, its beauty and its power to enthrall."

   —The New York Times

"Two fascinating narratives are interwoven in *Love and Math*, one mathematical, the other personal... Frenkel deftly takes the reader... to the far reaches of our current understanding. He seeks to lay bare the beauty of mathematics for everyone. As he writes, 'There is nothing in this world that is so deep and exquisite and yet so readily available to all.' "

   —Nature

"Frenkel has done an extraordinary job of making his case for love and mathematics. I think a lot of nonmathematicians will gain appreciation for the field, in the way that Stephen Hawking's *A Brief History of Time* delivered cutting-edge cosmology to the masses. It's not just the clarity of the thought or the skillful writing; in both cases, one of the best practitioners in the world has opened himself up personally to communicate deep ideas."

   —Wilmott

"Part ode, part autobiography, *Love and Math* is an admirable attempt to lay bare the beauty of numbers for all to see."

   —Scientific American

"Edward Frenkel mounts a passionate case against math's reputation as an arcane and boring field [and] argues for math's beauty and relevance."

   —Page-Turner blog, *The New Yorker*

"Frenkel pares the technical details to a minimum as he reflects on the platonic transcendence of mathematical concepts and marvels at their mysterious utility in explaining physical phenomena. Not merely dry formulas in textbooks, the math Frenkel celebrates fosters freedom and, yes, even distills the essence of love. A breathtaking personal and intellectual odyssey."

   —Booklist

"Fascinating... By using analogies, [Frenkel] describes concepts such as symmetries, dimensions, and Riemann surfaces in a way that will enable non-mathematicians to understand them. Whether or not readers develop a love for math, they will get a glimpse of the love that Frenkel has for the subject. Recommended for all readers, math whizzes or not, inclined to be interested in the subject."

—*Library Journal*

"Frenkel reveals the joy of pure intellectual discovery in this autobiographical story of determination, passion, and the Langlands program... Frenkel's gusto will draw readers into his own quest, pursuing the deepest realities of mathematics as if it were 'a giant jigsaw puzzle, in which no one knows what the final image is going to look like.' "

—*Publishers Weekly*

"A fascinating peek into the author's life and work."

—*Kirkus Reviews*

"Edward Frenkel's riveting new book made this former math phobe fall for a subject I thought I hated. He worships math with a passion so contagious, you'll be swept away – by both the subject and Frenkel's remarkable journey. When the USSR tried to block him from university based on his Jewish surname, he literally scaled twenty-foot fences to steal into classes. Before age 21, a letter from Harvard invited him to teach, launching a career that includes writing and starring in an erotic film paying homage to Yukio Mishima, *Rites of Love and Math*. Frenkel's charisma is undeniable. Not since G.H. Hardy's *Mathematician's Apology* has one of the field's finest minds clarified the metaphysical beauty of this misunderstood field of inquiry. This book is not just a love song for a subject and a battle cry for educational reform, it's literary pleasure at its freshest."

—**Mary Karr, bestselling author of *The Liars' Club*, *Cherry*, and *Lit***

"Love and Math = fast-paced adventure story + intimate memoir + insider's account of the quest to decode a Rosetta Stone at the heart of modern math. It all adds up to a thrilling intellectual ride – and a tale of surprising passion."

—**Steven Strogatz, Professor of Applied Mathematics, Cornell University, and author of *The Joy of x***

"I don't know if I've ever used the words love and math together, but this book changed that. Edward Frenkel writes of the objective beauty of numbers. Like musical notes, they exist apart from the mind, daring us to fathom their depths and assemble them in arcane narratives that tell the story of us. Reading this book, one is compelled to drop everything and give math another try; to partake of the ultimate mystery."

—Chris Carter, creator of *The X-Files*

"This very readable, passionately written, account of some of the most exciting ideas in modern mathematics is highly recommended to all who are curious lovers of beauty."

—David Gross, Nobel Laureate in Physics

"A marvelous and arresting account of the struggles, the joys, the passions of a mathematician. In this thrilling account of how Frenkel overcame the bleak anti-Semitism in his early schooling in Moscow to contribute to the grand goals of his subject, he makes the palette of mathematical ideas vivid to us by calling upon things as diverse as his mother's recipe for borscht (to explain the flavor of quantum duality) and imagined screenplays (to offer hints of the Langlands Program)."

—Barry Mazur, University Professor, Harvard University, and author of *Imagining Numbers*

"While you might think of Edward Frenkel as that mathematician who made that erotic film about math, actually you should know him as the guy who's going to help you see through your anxieties and perceive your world more deeply. *Love and Math* is an autobiography, a portal to understanding previously fearsome math, and the first popular account of the Langlands Program, which is one of the central creative projects of humanity at this time. This book is about knowing reality as fundamentally as possible on every level."

—Jaron Lanier, author of *You Are Not a Gadget* and *Who Owns the Future?*

"Through his fascinating autobiography, mathematician Edward Frenkel is opening for us a window into the ambitious Langlands Program – a sweeping network that interconnects many branches of mathematics and physics. A breathtaking view of modern mathematics."

—Mario Livio, astrophysicist, and author of *The Golden Ratio* and *Brilliant Blunders*

Love and Math

LOVE and MATH

The Heart of Hidden Reality

Edward Frenkel

BASIC BOOKS

A Member of the Perseus Books Group
New York

Hardcover first published in 2013 by Basic Books, A Member of the Perseus Books Group
Paperback first published in 2014 by Basic Books

The Library of Congress has cataloged the hardcover edition as follows:

Frenkel, Edward, 1968– author.
Love and math : the heart of hidden reality / Edward Frenkel.
    pages cm
  Includes bibliographical references and index.
  ISBN 978-0-465-05074-1 (hardback) – ISBN 978-0-465-06995-8 (e-book)
  1. Frenkel, Edward, 1968– 2. Mathematicians–United States–Biography.
  3. Mathematics–Miscellanea. I. Title.

QA29.F725F74 2013
510.92–dc23
[B]

                    2013017372

ISBN 978-0-465-06495-3 (paperback)

10 9 8 7 6 5 4 3 2 1

For my parents

Contents

Preface

There's a secret world out there. A hidden parallel universe of beauty and elegance, intricately intertwined with ours. It's the world of mathematics. And it's invisible to most of us. This book is an invitation to discover this world.

Consider this paradox: On the one hand, mathematics is woven in the very fabric of our daily lives. Every time we make an online purchase, send a text message, do a search on the Internet, or use a GPS device, mathematical formulas and algorithms are at play. On the other hand, most people are daunted by math. It has become, in the words of poet Hans Magnus Enzensberger, "a blind spot in our culture – alien territory, in which only the elite, the initiated few have managed to entrench themselves." It's rare, he says, that we "encounter a person who asserts vehemently that the mere thought of reading a novel, or looking at a picture, or seeing a movie causes him insufferable torment," but "sensible, educated people" often say "with a remarkable blend of defiance and pride" that math is "pure torture" or a "nightmare" that "turns them off."

How is this anomaly possible? I see two main reasons. First, mathematics is more abstract than other subjects, hence not as accessible. Second, what we study in school is only a tiny part of math, much of it established more than a millennium ago. Mathematics has advanced tremendously since then, but the treasures of modern math have been kept hidden from most of us.

What if at school you had to take an "art class" in which you were only taught how to paint a fence? What if you were never shown the paintings of Leonardo da Vinci and Picasso? Would that make you appreciate art? Would you want to learn more about it? I doubt it. You would probably say something like this: "Learning art at school was a waste of my time. If I ever need to have my fence painted, I'll just hire people to do this for me." Of course, this sounds ridiculous, but this is how math is taught, and so in the eyes of most of us it becomes the equivalent of watching paint dry. While the paintings of the great masters are readily available, the math of the great masters is locked away.

However, it's not just the aesthetic beauty of math that's captivating. As Galileo famously said, "The laws of Nature are written in the language of mathematics." Math is a way to describe reality and figure out how the world works, a universal language that has become the gold standard of truth. In our world, increasingly driven by science and technology, mathematics is becoming, ever more, the source of power, wealth, and progress. Hence those who are fluent in this new language will be on the cutting edge of progress.

One of the common misconceptions about mathematics is that it can only be used as a "toolkit": a biologist, say, would do some field work, collect data, and then try to build a mathematical model fitting these data (perhaps, with some help from a mathematician). While this is an important mode of operation, math offers us *a lot more*: it enables us to make groundbreaking, paradigm-shifting leaps that we couldn't make otherwise. For example, Albert Einstein was not trying to fit any data into equations when he understood that gravity causes our space to curve. In fact, there was no such data. No one could even imagine at the time that our space is curved; everyone "knew" that our world was flat! But Einstein understood that this was the only way to generalize his special relativity theory to non-inertial systems, coupled with his insight that gravity and acceleration have the same effect. This was a high-level intellectual exercise within the realm of math, one in which Einstein relied on the work of a mathematician, Bernhard Riemann, completed fifty years earlier. The human brain is wired in such a way that we simply cannot imagine curved spaces of dimension greater than two; we can only access them through mathematics. And guess what, Einstein was right – our universe *is* curved, and furthermore, it's expanding. That's the power of mathematics I am talking about!

Many examples like this may be found, and not only in physics, but in other areas of science (we will discuss some of them below). History shows that science and technology are transformed by mathematical ideas at an accelerated pace; even mathematical theories that are initially viewed as abstract and esoteric later become indispensable for applications. Charles Darwin, whose work at first did not rely on math, later wrote in his autobiography: "I have deeply regretted that I did not proceed far enough at least to understand something of the great leading principles of mathematics, for men thus endowed seem to have an extra sense." I take it as prescient advice to the next generations to capitalize on mathematics' immense potential.

When I was growing up, I wasn't aware of the hidden world of mathematics. Like most people, I thought math was a stale, boring subject. But I was lucky:

in my last year of high school I met a professional mathematician who opened the magical world of math to me. I learned that mathematics is full of infinite possibilities as well as elegance and beauty, just like poetry, art, and music. I fell in love with math.

Dear reader, with this book I want to do for you what my teachers and mentors did for me: unlock the power and beauty of mathematics, and enable *you* to enter this magical world the way I did, even if you are the sort of person who has never used the words "math" and "love" in the same sentence. Mathematics will get under your skin just like it did under mine, and your worldview will never be the same.

* * *

Mathematical knowledge is unlike any other knowledge. While our perception of the physical world can always be distorted, our perception of mathematical truths can't be. They are objective, persistent, necessary truths. A mathematical formula or theorem means the same thing to anyone anywhere – no matter what gender, religion, or skin color; it will mean the same thing to anyone a thousand years from now. And what's also amazing is that we own all of them. No one can patent a mathematical formula, it's ours to share. There is nothing in this world that is so deep and exquisite and yet so readily available to all. That such a reservoir of knowledge really exists is nearly unbelievable. It's too precious to be given away to the "initiated few." It belongs to all of us.

One of the key functions of mathematics is the ordering of information. This is what distinguishes the brush strokes of Van Gogh from a mere blob of paint. With the advent of 3D printing, the reality we are used to is undergoing a radical transformation: everything is migrating from the sphere of physical objects to the sphere of information and data. We will soon be able to convert information into matter on demand by using 3D printers just as easily as we now convert a PDF file into a book or an MP3 file into a piece of music. In this brave new world, the role of mathematics will become even more central: as the way to organize and order information, and as the means to facilitate the conversion of information into physical reality.

In this book, I will describe one of the biggest ideas to come out of mathematics in the last fifty years: the Langlands Program, considered by many as the Grand Unified Theory of mathematics. It's a fascinating theory that weaves a web of tantalizing connections between mathematical fields that at first glance seem to be light years apart: algebra, geometry, number theory, analysis, and quantum physics. If we think of those fields as continents in the hidden world of mathematics, then the Langlands Program is the ultimate teleportation device, capable of getting us instantly from one of them to another, and back.

Launched in the late 1960s by Robert Langlands, the mathematician who currently occupies Albert Einstein's office at the Institute for Advanced Study in Princeton, the Langlands Program had its roots in a groundbreaking mathematical theory of symmetry. Its foundations were laid two centuries ago by a French prodigy, just before he was killed in a duel, at age twenty. It was subsequently enriched by another stunning discovery, which not only led to the proof of Fermat's Last Theorem, but revolutionized the way we think about numbers and equations. Yet another penetrating insight was that mathematics has its own Rosetta stone and is full of mysterious analogies and metaphors. Following these analogies as creeks in the enchanted land of math, the ideas of the Langlands Program spilled into the realms of geometry and quantum physics, creating order and harmony out of seeming chaos.

I want to tell you about all this to expose the sides of mathematics we rarely get to see: inspiration, profound ideas, startling revelations. Mathematics is a way to break the barriers of the conventional, an expression of unbounded imagination in the search for truth. Georg Cantor, creator of the theory of infinity, wrote: "The essence of mathematics lies in its freedom." Mathematics teaches us to rigorously analyze reality, study the facts, follow them wherever they lead. It liberates us from dogmas and prejudice, nurtures the capacity for innovation. It thus provides tools that transcend the subject itself.

These tools can be used for good and for ill, forcing us to reckon with math's real-world effects. For example, the global economic crisis was caused to a large extent by the widespread use of inadequate mathematical models in the financial markets. Many of the decision makers didn't fully understand these models due to their mathematical illiteracy, but were arrogantly using them anyway – driven by greed – until this practice almost wrecked the entire system. They were taking unfair advantage of the asymmetric access to information and hoping that no one would call their bluff because others weren't inclined to ask how these mathematical models worked either. Perhaps, if more people understood how these models functioned, how the system really worked, we wouldn't have been fooled for so long.

As another example, consider this: in 1996, a commission appointed by the U.S. government gathered in secret and altered a formula for the Consumer

Price Index, the measure of inflation that determines the tax brackets, Social Security, Medicare, and other indexed payments. Tens of millions of Americans were affected, but there was little public discussion of the new formula and its consequences. And recently there was another attempt to exploit this arcane formula as a backdoor on the U.S. economy.[1]

Far fewer of these sorts of backroom deals could be made in a mathematically literate society. Mathematics equals rigor plus intellectual integrity times reliance on facts. We should all have access to the mathematical knowledge and tools needed to protect us from arbitrary decisions made by the powerful few in an increasingly math-driven world. Where there is no mathematics, there is no freedom.

* * *

Mathematics is as much part of our cultural heritage as art, literature, and music. As humans, we have a hunger to discover something new, reach new meaning, understand better the universe and our place in it. Alas, we can't discover a new continent like Columbus or be the first to set foot on the Moon. But what if I told you that you don't have to sail across an ocean or fly into space to discover the wonders of the world? They are right here, intertwined with our present reality. In a sense, within us. Mathematics directs the flow of the universe, lurks behind its shapes and curves, holds the reins of everything from tiny atoms to the biggest stars.

This book is an invitation to this rich and dazzling world. I wrote it for readers without any background in mathematics. If you think that math is hard, that you won't get it, if you are terrified by math, but at the same time curious whether there is something there worth knowing – then this book is for you.

There is a common fallacy that one has to study mathematics for years to appreciate it. Some even think that most people have an innate learning disability when it comes to math. I disagree: most of us have heard of and have at least a rudimentary understanding of such concepts as the solar system, atoms and elementary particles, the double helix of DNA, and much more, without taking courses in physics and biology. And nobody is surprised that these sophisticated ideas are part of our culture, our collective consciousness. Likewise, everybody can grasp key mathematical concepts and ideas, if they are explained in the right way. To do this, it is not necessary to study math for years; in many cases, we can cut right to the point and jump over tedious steps.

The problem is: while the world at large is always talking about planets, atoms, and DNA, chances are no one has ever talked to you about the fascinating ideas of modern math, such as symmetry groups, novel numerical systems in which 2 and 2 isn't always 4, and beautiful geometric shapes like Riemann surfaces. It's like they keep showing you a little cat and telling you that this is what a tiger looks like. But actually the tiger is an entirely different animal. I'll show it to you in all of its splendor, and you'll be able to appreciate its "fearful symmetry," as William Blake eloquently said.

Don't get me wrong: reading this book won't by itself make you a mathematician. Nor am I advocating that everyone should become a mathematician. Think about it this way: learning a small number of chords will enable you to play quite a few songs on a guitar. It won't make you the world's best guitar player, but it will enrich your life. In this book I will show you the chords of modern math, which have been hidden from you. And I promise that this will enrich your life.

One of my teachers, the great Israel Gelfand, used to say: "People think they don't understand math, but it's all about how you explain it to them. If you ask a drunkard what number is larger, 2/3 or 3/5, he won't be able to tell you. But if you rephrase the question: what is better, 2 bottles of vodka for 3 people or 3 bottles of vodka for 5 people, he will tell you right away: 2 bottles for 3 people, of course."

My goal is to explain this stuff to you in terms that you will understand.

I will also talk about my experience of growing up in the former Soviet Union, where mathematics became an outpost of freedom in the face of an oppressive regime. I was denied entrance to Moscow State University because of the discriminatory policies of the Soviet Union. The doors were slammed shut in front of me. I was an outcast. But I didn't give up. I would sneak into the University to attend lectures and seminars. I would read math books on my own, sometimes late at night. And in the end, I was able to hack the system. They didn't let me in through the front door; I flew in through a window. When you are in love, who can stop you?

Two brilliant mathematicians took me under their wings and became my mentors. With their guidance, I started doing mathematical research. I was still a college student, but I was already pushing the boundaries of the unknown. This was the most exciting time of my life, and I did it even though I was sure that the discriminatory policies would never allow me to have a job as a mathematician in the Soviet Union.

But there was a surprise in store: my first mathematical papers were smuggled abroad and became known, and I got invited to Harvard University as a Visiting Professor at age twenty-one. Miraculously, at exactly the same time *perestroika* in the Soviet Union lifted the iron curtain, and citizens were allowed to travel abroad. So there I was, a Harvard professor without a Ph.D., hacking the system once again. I continued on my academic path, which led me to research on the frontiers of the Langlands Program and enabled me to participate in some of the major advances in this area during the last twenty years. In what follows, I will describe spectacular results obtained by brilliant scientists as well as what happened behind the scenes.

This book is also about love. Once, I had a vision of a mathematician discovering the "formula of love," and this became the premise of a film *Rites of Love and Math*, which I will talk about later in the book. Whenever I show the film, someone always asks: "Does a formula of love really exist?"

My response: "Every formula we discover is a formula of love." Mathematics is the source of timeless profound knowledge, which goes to the heart of all matter and unites us across cultures, continents, and centuries. My dream is that all of us will be able to see, appreciate, and marvel at the magic beauty and exquisite harmony of these ideas, formulas, and equations, for this will give so much more meaning to our love for this world and for each other.

A Guide for the Reader

I have made every effort to present mathematical concepts in this book in the most elementary and intuitive way. However, I realize that some parts of the book are somewhat heavier on math (particularly, some parts of Chapters 8, 14, 15, and 17). *It is perfectly fine to skip* those parts that look confusing or tedious at the first reading. Coming back to those parts later, equipped with newly gained knowledge, you might find the material easier to follow. But that is usually not necessary in order to be able to follow what comes next.

Perhaps, a bigger point is that *it is perfectly OK if something is unclear*. That's how I feel 90 percent of the time when I do mathematics, so welcome to my world! The feeling of confusion (even frustration, sometimes) is an essential part of being a mathematician. But look at the bright side: how boring would life be if everything in it could be understood with little effort! What makes doing mathematics so exciting is our desire to overcome this confusion; to understand; to lift the veil on the unknown. And the feeling of personal triumph when we do understand something makes it all worthwhile.

My focus in this book is on the big picture and the logical connections between different concepts and different branches of math, not technical details. A more in-depth discussion is often relegated to the endnotes, which also contain references and suggestions for further reading. However, although endnotes may enhance your understanding, they may be safely skipped (at least, at the first reading).

I have tried to minimize the use of formulas – opting, whenever possible, for verbal explanations. Feel free to skip the few formulas that do appear.

A word of warning on mathematical terminology: while writing this book, I discovered, to my surprise, that certain terms that mathematicians use in a specific way actually mean something entirely different to non-mathematicians. Terms like correspondence, representation, composition, loop, manifold, and theory. Whenever I detected this issue, I included an explanation. Also, whenever possible, I changed obscure mathematical terms to terms with more transparent meaning (for example, I would write "Langlands relation" instead of "Langlands correspondence"). You might find it useful to consult the Glossary and the Index whenever there is a word that seems unclear.

Please check out my website http://edwardfrenkel.com for updates and supporting materials, and send me an e-mail to share your thoughts about the book (my e-mail address can be found on the website). Your feedback will be much appreciated.

Chapter 1

A Mysterious Beast

How does one become a mathematician? There are many ways that this can happen. Let me tell you how it happened to me.

It might surprise you, but I hated math when I was at school. Well, "hated" is perhaps too strong a word. Let's just say I didn't like it. I thought it was boring. I could do my work, sure, but I didn't understand why I was doing it. The material we discussed in class seemed pointless, irrelevant. What really excited me was physics – especially quantum physics. I devoured every popular book on the subject that I could get my hands on. I grew up in Russia, where such books were easy to find.

I was fascinated with the quantum world. Ever since ancient times, scientists and philosophers had dreamed about describing the fundamental nature of the universe – some even hypothesized that all matter consists of tiny pieces called atoms. Atoms were proved to exist at the beginning of the twentieth century, but at around the same time, scientists discovered that each atom could be divided further. Each atom, it turned out, consists of a nucleus in the middle and electrons orbiting it. The nucleus, in turn, consists of protons and neutrons, as shown on the diagram below.[1]

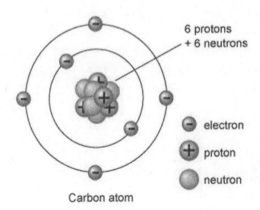

6 protons
+ 6 neutrons

⊖ electron

⊕ proton

⬤ neutron

Carbon atom

And what about protons and neutrons? The popular books that I was reading told me that they are built of the elementary particles called "quarks."

I liked the name quarks, and I especially liked how this name came about.
The physicist who invented these particles, Murray Gell-Mann, borrowed this
name from James Joyce's book *Finnegans Wake*, where there is a mock poem
that goes like this:

> Three quarks for Muster Mark!
> Sure he hasn't got much of a bark
> And sure any he has it's all beside the mark.

I thought it was really cool that a physicist would name a particle after
a novel. Especially such a complex and non-trivial one as *Finnegans Wake*.
I must have been around thirteen, but I already knew by then that scientists were supposed to be these reclusive and unworldly creatures who were
so deeply involved in their work that they had no interest whatsoever in other
aspects of life such as Art and Humanities. I wasn't like this. I had many
friends, liked to read, and was interested in many things besides science. I
liked to play soccer and spent endless hours chasing the ball with my friends.
I discovered Impressionist paintings around the same time (it started with a
big volume about Impressionism, which I found in my parents' library). Van
Gogh was my favorite. Enchanted by his works, I even tried to paint myself.
All of these interests had actually made me doubt whether I was really cut
out to be a scientist. So when I read that Gell-Mann, a great physicist, Nobel
Prize–winner, had such diverse interests (not only literature, but also linguis-
tics, archaeology, and more), I was very happy.

According to Gell-Mann, there are two different types of quarks, "up" and "down," and different mixtures of them give neutrons and protons their characteristics. A neutron is made of two down and one up quarks, and a proton is made of two up and one down quarks, as shown on the pictures.[2]

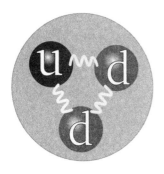 

Neutron   Proton

That was clear enough. But how physicists guessed that protons and neutrons were not indivisible particles but rather were built from smaller blocks was murky.

The story goes that by the late 1950s, a large number of apparently elementary particles, called hadrons, was discovered. Neutrons and protons are both hadrons, and of course they play major roles in everyday life as the building blocks of matter. As for the rest of hadrons – well, no one had any idea what they existed for (or "who ordered them," as one researcher put it). There were so many of them that the influential physicist Wolfgang Pauli joked that physics was turning into botany. Physicists desperately needed to rein in the hadrons, to find the underlying principles that govern their behavior and would explain their maddening proliferation.

Gell-Mann, and independently Yuval Ne'eman, proposed a novel classification scheme. They both showed that hadrons can be naturally split into small families, each consisting of eight or ten particles. They called them octets and decuplets. Particles within each of the families had similar properties.

In the popular books I was reading at the time, I would find octet diagrams like this:

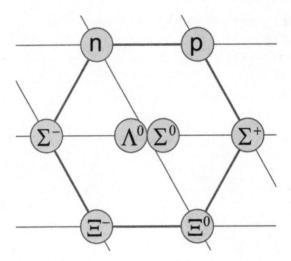

Here the proton is marked as p, the neutron is marked as n, and there are six other particles with strange names expressed by Greek letters.

But why 8 and 10, and not 7 and 11, say? I couldn't find a coherent explanation in the books I was reading. They would mention a mysterious idea of Gell-Mann called the "eightfold way" (referencing the "Noble Eightfold Path" of Buddha). But they never attempted to explain what this was all about.

This lack of explanation left me deeply unsatisfied. The key parts of the story remained hidden. I wanted to unravel this mystery but did not know how.

As luck would have it, I got help from a family friend. I grew up in a small industrial town called Kolomna, population 150,000, which was about seventy miles away from Moscow, or just over two hours by train. My parents worked as engineers at a large company making heavy machinery. Kolomna is an old town on the intersection of two rivers that was founded in 1177 (only thirty years after the founding of Moscow). There are still a few pretty churches and the city wall to attest to Kolomna's storied past. But it's not exactly an educational or intellectual center. There was only one small college there, which prepared schoolteachers. One of the professors there, a mathematician named Evgeny Evgenievich Petrov, however, was an old friend of my parents. And one day my mother met him on the street after a long time, and they started talking. My mom liked to tell her friends about me, so I came up in conversation. Hearing that I was interested in science, Evgeny Evgenievich said, "I must meet him. I will try to convert him to math."

"Oh no," my mom said, "he doesn't like math. He thinks it's boring. He wants to do quantum physics."

"No worries," replied Evgeny Evgenievich, "I think I know how to change his mind."

A meeting was arranged. I wasn't particularly enthusiastic about it, but I went to see Evgeny Evgenievich at his office anyway.

I was just about to turn fifteen, and I was finishing the ninth grade, the penultimate year of high school. (I was a year younger than my classmates because I had skipped the sixth grade.) Then in his early forties, Evgeny Evgenievich was friendly and unassuming. Bespectacled, with a beard stubble, he was just what I imagined a mathematician would look like, and yet there was something captivating in the probing gaze of his big eyes. They exuded unbounded curiosity about everything.

It turned out that Evgeny Evgenievich indeed had a clever plan how to convert me to math. As soon as I came to his office, he asked me, "So, I hear you like quantum physics. Have you heard about Gell-Mann's eightfold way and the quark model?"

"Yes, I've read about this in several popular books."

"But do you know what was the basis for this model? How did he come up with these ideas?"

"Well..."

"Have you heard about the group $SU(3)$?"

"SU what?"

"How can you possibly understand the quark model if you don't know what the group $SU(3)$ is?"

He pulled out a couple of books from his bookshelf, opened them, and showed me pages of formulas. I could see the familiar octet diagrams, such as the one shown above, but these diagrams weren't just pretty pictures; they were part of what looked like a coherent and detailed explanation.

Though I could make neither head nor tail of these formulas, it became clear to me right away that they contained the answers I had been searching for. This was a moment of epiphany. I was mesmerized by what I was seeing and hearing; touched by something I had never experienced before; unable to express it in words but feeling the energy, the excitement one feels from hearing a piece of music or seeing a painting that makes an unforgettable impression. All I could think was "Wow!"

"You probably thought that mathematics is what they teach you in school,"

Evgeny Evgenievich said. He shook his head, "No, this" – he pointed at the formulas in the book – "is what mathematics is about. And if you really want to understand quantum physics, this is where you need to start. Gell-Mann predicted quarks using a beautiful mathematical theory. It was in fact a mathematical discovery."

"But how do I even begin to understand this stuff?"

It looked kind of scary.

"No worries. The first thing you need to learn is the concept of a symmetry group. That's the main idea. A large part of mathematics, as well as theoretical physics, is based on it. Here are some books I want to give you. Start reading them and mark the sentences that you don't understand. We can meet here every week and talk about this."

He gave me a book about symmetry groups and also a couple of others on different topics: about the so-called *p*-adic numbers (a number system radically different from the numbers we are used to) and about topology (the study of the most fundamental properties of geometric shapes). Evgeny Evgenievich had impeccable taste: he found a perfect combination of topics that would allow me to see this mysterious beast – *Mathematics* – from different sides and get excited about it.

At school we studied things like quadratic equations, a bit of calculus, some basic Euclidean geometry, and trigonometry. I had assumed that all mathematics somehow revolved around these subjects, that perhaps problems became more complicated but stayed within the same general framework I was familiar with. But the books Evgeny Evgenievich gave me contained glimpses of an entirely different world, whose existence I couldn't even imagine.

I was instantly converted.

Chapter 2

The Essence of Symmetry

In the minds of most people, mathematics is all about numbers. They imagine mathematicians as people who spend their days crunching numbers: big numbers, and even bigger numbers, all having exotic names. I had thought so too – at least, until Evgeny Evgenievich introduced me to the concepts and ideas of modern math. One of them turned out to be the key to the discovery of quarks: the concept of symmetry.

What is symmetry? All of us have an intuitive understanding of it – we know it when we see it. When I ask people to give me an example of a symmetric object, they point to butterflies, snowflakes, or the human body.

Photo by K.G. Libbrecht

But if I ask them what we mean when we say that an object is symmetrical, they hesitate.

Here is how Evgeny Evgenievich explained it to me. "Let's look at this round table and this square table," he pointed at the two tables in his office. "Which one is more symmetrical?"

"Of course, the round table, isn't it obvious?"

"But why? Being a mathematician means that you don't take 'obvious' things for granted but try to reason. Very often you'll be surprised that the most obvious answer is actually wrong."

Noticing confusion on my face, Evgeny Evgenievich gave me a hint: "What is the property of the round table that makes it more symmetrical?"

I thought about this for a while, and then it hit me: "I guess the symmetry of an object has to do with it keeping its shape and position unchanged even when we apply changes to it."

Evgeny Evgenievich nodded.

"Indeed. Let's look at all possible transformations of the two tables which preserve their shape and position," he said. "In the case of the round table..."

I interrupted him: "Any rotation around the center point will do. We will get back the same table positioned in the same way. But if we apply an arbitrary rotation to a square table, we will typically get a table positioned differently. Only rotations by 90 degrees and its multiples will preserve it."

"Exactly! If you leave my office for a minute, and I turn the round table by any angle, you won't notice the difference. But if I do the same to the square table, you will, unless I turn it by 90, 180, or 270 degrees."

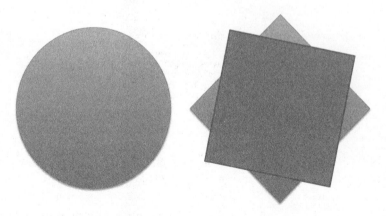

Rotation of a round table by any angle does not change its position, but rotation of a square table by an angle that is not a multiple of 90 degrees does change its position (both are viewed here from above)

He continued: "Such transformations are called symmetries. So you see that the square table has only four symmetries, whereas the round table has many more of them – it actually has infinitely many symmetries. That's why we say that the round table is more symmetrical."

This made a lot of sense.

"This is a fairly straightforward observation," continued Evgeny Evgeni-evich. "You don't have to be a mathematician to see this. But if you are a mathematician, you ask the next question: what are *all* possible symmetries of a given object?"

Let's look at the square table. Its symmetries[1] are these four rotations around the center of the table: by 90 degrees, 180 degrees, 270 degrees, and 360 degrees, counterclockwise.[2] A mathematician would say that the *set* of symmetries of the square table consists of four elements, corresponding to the angles 90, 180, 270, and 360 degrees. Each rotation takes a fixed corner (marked with a balloon on the picture below) to one of the four corners.

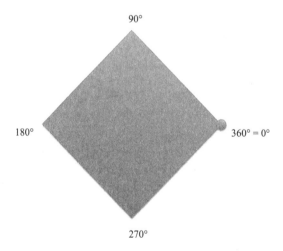

One of these rotations is special; namely, rotation by 360 degrees is the same as rotation by 0 degrees, that is, no rotation at all. This is a special symmetry because it actually does nothing to our object: each point of the table ends up in exactly the same position as it was before. We call it the *identical symmetry*, or just the *identity*.[3]

Note that rotation by any angle greater than 360 degrees is equivalent to rotation by an angle between 0 and 360 degrees. For example, rotation by 450 degrees is the same as rotation by 90 degrees, because 450 = 360 + 90. That's why we will only consider rotations by angles between 0 and 360 degrees.

Here comes the crucial observation: if we apply two rotations from the list $\{90°, 180°, 270°, 360°\}$ one after another, we obtain another rotation from the same list. We call this new symmetry the *composition* of the two.

Of course, this is obvious: each of the two symmetries preserves the table. Hence the composition of the two symmetries also preserves it. Therefore this composition has to be a symmetry as well. For example, if we rotate the table by 90 degrees and then again by 180 degrees, the net result is the rotation by 270 degrees.

Let's see what happens with the table under these symmetries. Under the counterclockwise rotation by 90 degrees, the right corner of the table (the one marked with a balloon on the previous picture) will go to the upper corner. Next, we apply the rotation by 180 degrees, so the upper corner will go to the down corner. The net result will be that the right corner will go to the down corner. This is the result of the counterclockwise rotation by 270 degrees.

Here is one more example:

$$90° + 270° = 0°.$$

By rotating by 90 degrees and then by 270 degrees, we get the rotation by 360 degrees. But the effect of the rotation by 360 degrees is the same as that of the rotation by 0 degrees, as we have discussed above – this is the "identity symmetry."

In other words, the second rotation by 270 degrees undoes the initial rotation by 90 degrees. This is in fact an important property: any symmetry can be *undone*; that is, for any symmetry S there exists another symmetry S' such that their composition is the identity symmetry. This S' is called the *inverse* of symmetry S. So we see that rotation by 270 degrees is the inverse of the rotation by 90 degrees. Likewise, the inverse of the rotation by 180 degrees is the same rotation by 180 degrees.

We now see that what looks like a very simple collection of symmetries of the square table – the four rotations $\{90°, 180°, 270°, 0°\}$ – actually has a lot of inner structure, or rules for how the members of the set can interact.

First of all, we can compose any two symmetries (that is, apply them one after another).

Second, there is a special symmetry, the identity. In our example, this is the rotation by 0 degrees. If we compose it with any other symmetry, we get back the same symmetry. For example,

$$90° + 0° = 90°, \qquad 180° + 0° = 180°, \qquad \text{etc.}$$

Third, for any symmetry S, there is the inverse symmetry S' such that the composition of S and S' is the identity.

And now we come to the main point: the set of rotations along with these three structures comprise an example of what mathematicians call a *group*.

The symmetries of any other object also constitute a group, which in general has more elements – possibly, infinitely many.[4]

Let's see how this works in the case of a round table. Now that we have gained some experience, we can see right away that the set of all symmetries of the round table is just the set of all possible rotations (not just by multiples of 90 degrees), and we can visualize it as the set of all points of a circle.

Each point on this circle corresponds to an angle between 0 and 360 degrees, representing the rotation of the round table by this angle in the counterclockwise direction. In particular, there is a special point corresponding to rotation by 0 degrees. It is marked on the picture below, together with another point corresponding to rotation by 30 degrees.

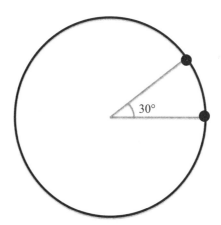

We should not think of the points of this circle as points of the round table, though. Rather, each point of the circle represents a particular rotation of the round table. Note that the round table does not have a preferred point, but our circle does; namely, the one corresponding to rotation by 0 degrees.[5]

Now let's see if the above three structures can be applied to the set of points of the circle.

First, the composition of two rotations, by φ_1 and φ_2 degrees, is the rotation by $\varphi_1 + \varphi_2$ degrees. If $\varphi_1 + \varphi_2$ is greater than 360, we simply subtract 360 from it. In mathematics, this is called *addition modulo* 360. For example, if $\varphi_1 = 195$ and $\varphi_2 = 250$, then the sum of the two angles is 445, and the rotation by 445 degrees is the same as the rotation by 85 degrees. So, in the group of rotations of the round table we have

$$195° + 250° = 85°.$$

Second, there is a special point on the circle corresponding to the rotation by 0 degrees. This is the identity element of our group.

Third, the inverse of the counterclockwise rotation by φ degrees is the counterclockwise rotation by $(360 - \varphi)$ degrees, or equivalently, clockwise rotation by φ degrees (see the drawing).

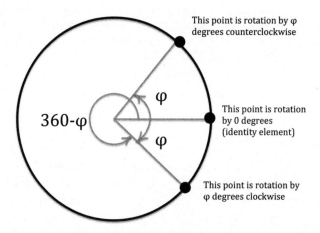

This point is rotation by φ degrees counterclockwise

This point is rotation by 0 degrees (identity element)

This point is rotation by φ degrees clockwise

Thus, we have described the group of rotations of the round table. We will call it the *circle group*. Unlike the group of symmetries of the square table, which has four elements, this group has infinitely many elements because there are infinitely many angles between 0 and 360 degrees.

We have now put our intuitive understanding of symmetry on firm theoretical ground – indeed, we've turned it into a mathematical concept. First, we postulated that a symmetry of a given object is a transformation that preserves it and its properties. Then we made a decisive step: we focused on the set of all symmetries of a given object. In the case of a square table, this set consists of four elements (rotations by multiples of 90 degrees); in the case of a round table, it is an infinite set (of all points on the circle). Finally, we described the neat structures that this set of symmetries always possesses: any two symmetries can be composed to produce another symmetry, there exists the identical symmetry, and for each symmetry there exists its inverse. (The composition of symmetries also satisfies the associativity property described in endnote 4.) Thus, we came to the mathematical concept of a group.

A group of symmetries is an abstract object that is quite different from the concrete object we started with. We cannot touch or hold the set of symmetries of a table (unlike the table itself), but we can imagine it, draw its elements, study it, talk about it. Each element of this abstract set has a concrete meaning, though: it represents a particular transformation of a concrete object, its symmetry.

Mathematics is about the study of such abstract objects and concepts.

Experience shows that symmetry is an essential guiding principle for the laws of nature. For example, a snowflake forms a perfect hexagonal shape because that turns out to be the lowest energy state into which crystallized water molecules are forced. The symmetries of the snowflake are rotations by multiples of 60 degrees; that is, 60, 120, 180, 240, 300, and 360 (which is the same as 0 degrees). In addition, we can "flip" the snowflake along each of the six axes corresponding to those angles. All of these rotations and flips preserve the shape and position of the snowflake, and hence they are its symmetries.*

In the case of a butterfly, flipping it turns it upside down. Since it has legs on one side, the flip is not, strictly speaking, a symmetry of the butterfly. When we say that a butterfly is symmetrical, we are talking about an idealized version of it, where its front and back are exactly the same (unlike those of an actual butterfly). Then the flip exchanging the left and the right wings becomes a symmetry. (Alternatively, we can imagine exchanging the wings without turning the butterfly upside down.)

This brings up an important point: there are many objects in nature whose symmetries are approximate. A real-life table is not perfectly round or per-

---

*Note that flipping a table is not a symmetry: this would turn it upside down – let's not forget that a table has legs. If we were to consider a square or a circle (no legs attached), then flips would be bona fide symmetries. We would have to include them in the corresponding symmetry groups.

fectly square, a live butterfly has an asymmetry between its front and back, and a human body is not fully symmetrical. However, even in this case it turns out to be useful to consider their abstract, idealized versions, or models – a perfectly round table or an image of the butterfly in which we don't distinguish between the front and the back. We then explore the symmetries of these idealized objects and adjust whatever inferences we can make from this analysis to account for the difference between a real-life object and its model.

This is not to say that we do not appreciate asymmetry; we do, and we often find beauty in it. But the main point of the mathematical theory of symmetry is not aesthetic. It is to formulate the concept of symmetry in the most general, and hence inevitably most abstract, terms so that it could be applied in a unified fashion in different domains, such as geometry, number theory, physics, chemistry, biology, and so on. Once we develop such a theory, we can also talk about the mechanisms of symmetry breaking – viewing asymmetry as emergent, if you will. For example, elementary particles acquire masses because the so-called gauge symmetry they obey (which will be discussed in Chapter 16) gets broken. This is facilitated by the Higgs boson, an elusive particle recently discovered at the Large Hadron Collider under the city of Geneva.[6] The study of such mechanisms of symmetry breaking yields invaluable insights into the behavior of the fundamental blocks of nature.

I'd like to point out some of the basic qualities of the abstract theory of symmetry because this is a good illustration of why mathematics is important.

The first is *universality*. The circle group is not only the group of symmetries of a round table, but also of all other round objects, like a glass, a bottle, a column, and so forth. In fact, to say that a given object is round is the same as to say that its group of symmetries is the circle group. This is a powerful statement: we realize that we can describe an important attribute of an object ("being round") by describing its symmetry group (the circle). Likewise, "being square" means that the group of symmetries is the group of four elements described above. In other words, the same abstract mathematical object (such as the circle group) serves many different concrete objects, and it points to universal properties that they all have in common (such as roundness).[7]

The second is *objectivity*. The concept of a group, for example, is independent of our interpretation. It means the same thing to anyone who learns it. Of course, in order to understand it, one has to know the language in which it is expressed, that is, mathematical language. But anyone can learn this language. Likewise, if you want to understand the meaning of René Descartes' sentence

"*Je pense, donc je suis*," you need to know French (at least, those words that are used in this sentence) – but anyone can learn it. However, in the case of the latter sentence, once we understand it, different interpretations of it are possible. Also, different people may agree or disagree on whether a particular interpretation of this sentence is true or false. In contrast, the meaning of a logically consistent mathematical statement is not subject to interpretation.[8] Furthermore, its truth is also objective. (In general, the truth of a particular statement may depend on the system of axioms within which it is considered. However, even then, this dependence on the axioms is also objective.) For example, the statement "the group of symmetries of a round table is a circle" is true to anyone, anywhere, at any time. In other words, mathematical truths are the necessary truths. We will talk more about this in Chapter 18.

The third, closely related, quality is *endurance*. There is little doubt that the Pythagorean theorem meant the same thing to the ancient Greeks as it does to us today, and there is every reason to expect that it will mean the same thing to anyone in the future. Likewise, all true mathematical statements we talk about in this book will remain true forever.

The fact that such objective and enduring knowledge exists (and moreover, belongs to all of us) is nothing short of a miracle. It suggests that mathematical concepts exist in a world separate from the physical and mental worlds – which is sometimes referred to as the Platonic world of mathematics (we will talk more about that in the closing chapter). We still don't fully understand what it is and what drives mathematical discovery. But it's clear that this hidden reality is bound to play a larger and larger role in our lives, especially with the advent of new computer technologies and 3D printing.

The fourth quality is *relevance* of mathematics to the physical world. For example, a lot of progress has been made in quantum physics in the past fifty years because of the application of the concept of symmetry to elementary particles and interactions between them. From this point of view, a particle, such as an electron or a quark, is like a round table or a snowflake, and its behavior is very much determined by its symmetries. (Some of these symmetries are exact, and some are approximate.)

The discovery of quarks is a perfect example of how this works. Reading the books Evgeny Evgenievich gave me, I learned that at the root of the Gell-Mann and Ne'eman classification of hadrons that we talked about in the previous chapter is a *symmetry group*. This group had been previously studied by mathematicians – who did not anticipate any connections to subatomic particles whatsoever. The mathematical name for it is $SU(3)$. Here S and U stand for "special unitary." This group is very similar in its properties to the group of symmetries of the sphere, which we will talk about in detail in Chapter 10.

Mathematicians had previously described the representations of the group $SU(3)$, that is, different ways that the group $SU(3)$ can be realized as a symmetry group. Gell-Mann and Ne'eman noticed the similarity between the structure of these representations and the patterns of hadrons that they had found. They used this information to classify hadrons.

The word "representation" is used in mathematics in a particular way, which is different from its more common usage. So let me pause and explain what this word means in the present context. Perhaps, it would help if I first give an example. Recall the group of rotations of a round table discussed above, the circle group. Now imagine extending the tabletop infinitely far in all directions. This way we obtain an abstract mathematical object: a plane. Each rotation of the tabletop, around its center, gives rise to a rotation of this plane around the same point. Thus, we obtain a rule that assigns a symmetry of this plane (a rotation) to every element of the circle group. In other words, each element of the circle group may be represented by a symmetry of the plane. For this reason mathematicians refer to this process as a *representation* of the circle group.

Now, the plane is two-dimensional because it has two coordinate axes and hence each point has two coordinates:

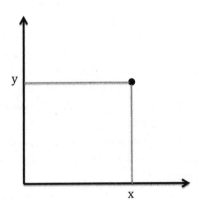

Therefore, we say that we have constructed a "two-dimensional representation" of the group of rotations. It simply means that each element of the group of rotations is realized as a symmetry of a plane.[9]

There are also spaces of dimension greater than two. For example, the space around us is three-dimensional. That is to say, it has three coordinate axes, and so in order to specify a position of a point, we need to specify its three coordinates (x,y,z) as shown on this picture:

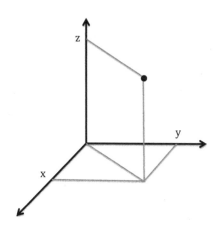

We cannot imagine a four-dimensional space, but mathematics gives us a universal language that allows us to talk about spaces of any dimension. Namely, we represent points of the four-dimensional space by quadruples of numbers (x,y,z,t), just like points of the three-dimensional space are represented by triples of numbers (x,y,z). In the same way, we represent points of an n-dimensional space, for any natural number n, by n-tuples of numbers. If you have used a spreadsheet program, then you have encountered such n-tuples: they appear as rows in a spreadsheet, each of the n numbers corresponding to a particular attribute of the stored data. Thus, every row in a spreadsheet refers to a point in an n-dimensional space. (We will talk more about spaces of various dimensions in Chapter 10.)

If each element of a group can be realized, in a consistent manner,[10] as a symmetry of an n-dimensional space, then we say that the group has an "n-dimensional representation."

It turns out that a given group can have representations of different dimensions. The reason elementary particles can be assembled in families of 8 and 10 particles is that the group $SU(3)$ is known to have an 8-dimensional and a 10-dimensional representation. The 8 particles of each octet constructed by Gell-Mann and Ne'eman (like the one shown on the diagram on p. 12) are in one-to-one correspondence with the 8 coordinate axes of an 8-dimensional space which is a representation of $SU(3)$. The same goes for the decuplet of particles. (But particles cannot be assembled in families of, say, 7 or 11 particles because mathematicians have proved that the group $SU(3)$ has no 7- or 11-dimensional representations.)

At first, this was just a convenient way to combine the particles with similar properties. But then Gell-Mann went further. He postulated that there was a deep reason behind this classification scheme. He basically said that this scheme works so well because hadrons consist of smaller particles – sometimes two and sometimes three of them – the quarks. A similar proposal was made independently by physicist George Zweig (who called the particles "aces").

This was a stunning proposal. Not only did it go against the popular belief at the time that protons and neutrons as well as other hadrons were indivisible elementary particles, these new particles were supposed to have electric charges that were fractions of the charge of the electron. This was a startling prediction because no one had seen such particles before. Yet, quarks were soon found experimentally, and as predicted, they had fractional electric charges!

What motivated Gell-Mann and Zweig to predict the existence of quarks? Mathematical theory of representations of the group $SU(3)$. Specifically, the fact that the group $SU(3)$ has two different 3-dimensional representations. (Actually, that's the reason there is a "3" in this group's name.) Gell-Mann and Zweig suggested that these two representations should describe two families of fundamental particles: 3 quarks and 3 anti-quarks. It turns out that the 8- and 10-dimensional representations of $SU(3)$ can be built from the 3-dimensional ones. And this gives us a precise blueprint for how to construct hadrons from quarks – just like in Lego.

Gell-Mann named the 3 quarks "up," "down," and "strange."[11] A proton consists of two up quarks and one down quark, whereas a neutron consists of two down quarks and one up quark, as we saw on the pictures on p. 11. Both of these particles belong to the octet shown on the diagram on p. 12. Other particles from this octet involve the strange quark as well as the up and down quarks. There are also octets that consist of particles that are composites of one quark and one anti-quark.

The discovery of quarks is a good example of the paramount role played by mathematics in science that we discussed in the Preface. These particles were predicted not on the basis of empirical data, but on the basis of mathematical symmetry patterns. This was a purely theoretical prediction, made within the framework of a sophisticated mathematical theory of representations of the group $SU(3)$. It took physicists years to master this theory (and in fact there was some resistance to it at first), but it is now the bread and butter of elementary particle physics. Not only did it provide a classification of hadrons, it also led to the discovery of quarks, which forever changed our understanding of physical reality.

Imagine: a seemingly esoteric mathematical theory empowered us to get to the heart of the building blocks of nature. How can we not be enthralled by the magic harmony of these tiny blobs of matter, not marvel at the capacity of mathematics to reveal the inner workings of the universe?

As the story goes, Albert Einstein's wife Elsa remarked, upon hearing that a telescope at the Mount Wilson Observatory was needed to determine the shape of space-time: "Oh, my husband does this on the back of an old envelope."

Physicists do need expensive and sophisticated machines such as the Large Hadron Collider in Geneva, but the amazing fact is that scientists like Einstein and Gell-Mann have used what looks like the purest and most abstract mathematical knowledge to unlock the deepest secrets of the world around us.

Regardless of who we are and what we believe in, we all share this knowledge. It brings us closer together and gives a new meaning to our love for the universe.

Chapter 3

The Fifth Problem

Evgeny Evgenievich's plan worked perfectly: I was "converted" to math. I was learning quickly, and the deeper I delved into math, the more my fascination grew, the more I wanted to know. This is what happens when you fall in love.

I started meeting with Evgeny Evgenievich on a regular basis. He would give me books to read, and I would meet him once a week at the pedagogical college where he taught to discuss what I had read. Evgeny Evgenievich played soccer, ice hockey, and volleyball on a regular basis, but like many men in the Soviet Union in those days, he was a chain smoker. For a long time afterward, the smell of cigarettes was associated in my mind with doing mathematics.

Sometimes our conversations would last well into the night. Once, the auditorium we were in was locked by the custodian who couldn't fathom that there would be someone inside at such a late hour. And we must have been so deep into our conversation that we didn't hear the turning of the key. Fortunately, the auditorium was on the ground floor, and we managed to escape through a window.

The year was 1984, my senior year at high school. I had to decide which university to apply to. Moscow had many schools, but there was only one place to study pure math: Moscow State University, known by its Russian abbreviation MGU, for *Moskovskiy Gosudarstvenny Universitet*. Its famous *Mekh-Mat*, the Department of Mechanics and Mathematics, was the flagship mathematics program of the USSR.

Entrance exams to colleges in Russia are not like the SAT that American students take. At *Mekh-Mat* there were four: written math, oral math, literature essay composition, and oral physics. Those who, like me, graduated from high school with highest honors (in the Soviet Union we were then given a gold medal) would be automatically accepted after getting a 5, the highest grade, at the first exam.

I had by then progressed far beyond high school math, and so it looked like I would sail through the exams at MGU.

But I was too optimistic. The first warning shot came in the form of a letter I received from a correspondence school with which I had studied. This school had been organized some years earlier by, among others, Israel Gelfand, the famous Soviet mathematician (we will talk much more about him later). The school intended to help those students who, like me, lived outside of major cities and did not have access to special mathematical schools. Every month, participating students would receive a brochure elucidating the material that was studied in school and going a little beyond. It also contained some problems, more difficult than those discussed at school, which a student was supposed to solve and mail back. Graders (usually undergrads of Moscow University) read those solutions and returned them, marked, to the students. I was enrolled in this school for three years, as well as in another school, which was more physics-oriented. It was a helpful resource for me, though the material was pretty close to what was studied at school (unlike the stuff I was studying privately with Evgeny Evgenievich).

The letter I received from this correspondence school was short: "If you would like to apply to Moscow University, stop by our office, and we will be happy to give you advice," and it gave the address on the campus of MGU and the visiting hours. Shortly after receiving the letter, I took the two-hour train ride to Moscow. The school's office was a big room with a bunch of desks and a number of people working, typing, and correcting papers. I introduced myself, produced my little letter, and was immediately led to a diminutive young woman, in her early thirties.

"What's your name?" she said by way of greeting.

"Eduard Frenkel." (I used the Russian version of Edward in those days.)

"And you want to apply to MGU?"

"Yes."

"Which department?"

"*Mekh-Mat.*"

"I see." She lowered her eyes and asked:

"And what's your nationality?"

I said, "Russian."

"Really? And what are your parents' nationalities?"

"Well... My mother is Russian."

"And your father?"

"My father is Jewish."

She nodded.

This dialogue might sound surreal to you, and as I am writing it now, it sounds surreal to me too. But in the Soviet Union circa 1984 – remember Orwell?* – it was not considered bizarre to ask someone what his or her "nationality" was. In the interior passport that all Soviet citizens had to carry with them, there was in fact a special line for "nationality." It came after (1) first name, (2) patronymic name, (3) last name, and (4) the date of birth. For this reason, it was called *pyataya grafa*, "the fifth line." Nationality was also recorded in one's birth certificate, as were the nationalities of the parents. If their nationalities were different, as in my case, then the parents had a choice of which nationality to give to their child.

For all intents and purposes, the fifth line was a code for asking whether one was Jewish or not. (People of other nationalities, such as Tatars and Armenians, against whom there were prejudices and persecution – though not nearly at the same scale as against the Jews – were also picked up this way.) My fifth line said that I was Russian, but my last name – which was my father's last name, and clearly sounded Jewish – gave me away.

It is important to note that my family was not religious at all. My father was not brought up in a religious tradition, and neither was I. Religion in the Soviet Union was in fact all but non-existent in those days. Most Christian Orthodox churches were destroyed or closed. In the few existing churches, one could typically only find a few old *babushkas* (grandmothers), such as my maternal grandmother. She occasionally attended service at the only active church in my hometown. There were even fewer synagogues. There were none in my hometown; in Moscow, whose population was close to 10 million, officially there was only one synagogue.[1] Going to a service in a church or a synagogue was dangerous: one could be spotted by special plain-clothed agents and would then get in a lot of trouble. So when someone was referred to as being Jewish, it was meant not in the sense of religion but rather in the sense of ethnicity, of "blood."

Even if I hadn't been using my father's last name, my Jewish origin would be picked up by the admissions committee anyway, because the application form specifically asked for the full names of both parents. Those full names

---

*This was one year before Mikhail Gorbachev came to power in the Soviet Union, and another couple of years before he launched his *perestroika*. The totalitarian Soviet regime in 1984 was in many ways a haunting facsimile of George Orwell's prescient book.

included patronymic names; that is, the first names of the grandparents of the applicant. My father's patronymic name is Joseph, which sounded unmistakably Jewish in the Soviet Union of that era, so this was another way to find out (if his last name hadn't given me away). The system was set up in such a way that it would flag those who were at least one-quarter Jewish.

Having established that by this definition I was a Jew, the woman said, "Do you know that Jews are not accepted to Moscow University?"

"What do you mean?"

"What I mean is that you shouldn't even bother to apply. Don't waste your time. They won't let you in."

I didn't know what to say.

"Is that why you sent me this letter?"

"Yes. I'm just trying to help you."

I looked around. It was clear that everyone in the office was aware of what this conversation was about, even if they weren't listening closely. This must have already happened dozens of times, and everybody seemed used to it. They all averted their eyes, as if I were a terminally ill patient. My heart sank.

I had encountered anti-Semitism before, but at a personal, not institutional, level. When I was in fifth grade, some of my classmates took to taunting me with *evrey, evrey* ("Jew, Jew"). I don't think they had any idea what this meant (which was clear from the fact that some of them confused the word *evrey*, which meant "Jew," with *evropeyets*, which meant "European") – they must have heard anti-Semitic remarks from their parents or other adults. (Unfortunately, anti-Semitism was deeply rooted in the Russian culture.) I was strong enough and lucky enough to have a couple of true friends who stood by me, so I was never actually beaten up by these bullies, but this was an unpleasant experience. I was too proud to tell the teachers or my parents, but one day a teacher overheard and intervened. As a result, those kids were immediately called to the principal, and the taunting stopped.

My parents had heard of the discrimination against Jews in entrance exams to universities, but somehow they did not pay much attention to this. In my hometown, there weren't many Jews to begin with, and all the purported discrimination cases my parents had heard of concerned programs in physics. A typical argument went that Jews weren't accepted there because the studies in such a program were related to nuclear research and hence to national defense and state secrets; the government didn't want Jews in those areas because Jews could emigrate to Israel or somewhere else. By this logic, there shouldn't have been a reason to care about those who studied pure math. Well, apparently, someone did.

Everything about my conversation at MGU was strange. And I am not just talking about the Kafkaesque aspect of it. It is possible to conclude that the woman I talked to simply tried to help me and other students by warning us of what's going to happen. But could this really be the case? Remember, we are talking about 1984, when the Communist Party and the KGB still tightly controlled all aspects of life in the Soviet Union. The official policy of the state was that all nationalities were equal, and publicly suggesting otherwise would put one in danger. Yet, this woman calmly talked about this to me, a stranger she had just met, and she didn't seem to be worried about being overheard by her colleagues.

Besides, the exams at MGU were always scheduled one month ahead of all other schools. Therefore, students who were failed at MGU would still have a chance to apply elsewhere. Why would someone try to convince them not even to try? It sounded like some powerful forces were trying to scare me and other Jewish students away.

But I would not be deterred. After talking about all this at great length, my parents and I felt that I had nothing to lose. We decided that I would apply to MGU anyway and just hope for the best.

The first exam, at the beginning of July, was a written test in mathematics. It always consisted of five problems. The fifth problem was considered deadly and unsolvable. It was like the fifth element of the exam. But I was able to solve all problems, including the fifth. Aware as I was of the strong likelihood that whoever graded my exam could be biased against me and would try to find gaps in my solutions, I wrote everything out in excruciating detail. I then checked and double-checked all my arguments and calculations to make sure that I hadn't made any mistakes. Everything looked perfect! I was in an upbeat mood on the train ride home. The next day I told Evgeny Evgenievich my solutions, and he confirmed that everything was correct. It seemed like I was off to a good start.

My next exam was oral math. It was scheduled for July 13, which happened to be a Friday.

I remember very clearly many details about that day. The exam was scheduled for the early afternoon, and I took the train from home with my mother that morning. I entered the room at MGU a few minutes before the exam. It was a regular classroom, and there were probably between fifteen and twenty students there and four or five examiners. At the start of the exam each of us had to draw a piece of paper from a big pile on the desk at the front of the room. Each paper had two questions written on it, and it was turned blank side up. It was like drawing a lottery ticket, so we called this piece of paper *bilet*, ticket. There were perhaps one hundred questions altogether, all known in advance. I didn't really care which ticket I would draw as I knew this material inside-out. After drawing the ticket, each student had to sit down at one of the desks and prepare the answer, using only the provided blank sheets of paper.

The questions on my ticket were: (1) a circle inscribed in a triangle and the formula for the area of the triangle using its radius; and (2) derivative of the ratio of two functions (the formula only). I was so ready for these questions, I could have answered them in my sleep.

I sat down, wrote down a few formulas on a sheet of paper, and collected my thoughts. This must have taken me about two minutes. There was no need to prepare more; I was ready. I raised my hand. There were several examiners present in the room, and they were all waiting for the students to raise their hands, but, bizarrely, they ignored me, as if I did not exist. They looked right through me. I was sitting with my hand raised for a while: no response.

Then, after ten minutes or so, a couple of other kids raised their hands, and as soon as they did, the examiners rushed to them. An examiner would take a seat next to a student and listen to him or her answer the questions. They were quite close to me, so I could hear them. The examiners were very polite and were mostly nodding their heads, only occasionally asking follow-up questions. Nothing out of the ordinary. When a student finished answering the questions on the ticket (after ten minutes or so), the examiner would give him or her one additional problem to solve. Those problems seemed rather simple, and most students solved them right away. And that was it!

The first couple of students were already happily gone, having obviously earned a 5, the highest grade, and I was still sitting there. Finally, I grabbed one of the examiners passing by, a young fellow who seemed like he was a fresh Ph.D., and asked him pointedly: "Why aren't you talking to me?" He looked away and said quietly: "Sorry, we are not allowed to talk to you."

An hour or so into the exam, two middle-aged men entered the room. They briskly walked up to the table at the front of the room and presented themselves to the guy who was sitting there. He nodded and pointed at me. It became clear that these were the people I'd been waiting for: my inquisitors.

They came up to my desk and introduced themselves. One was lean and quick, the other slightly overweight and with a big mustache.

"OK," the lean man said – he did most of the talking – "what have we got here? What's the first question?"

"The circle inscribed in a triangle and... "

He interrupted me: "What is the definition of a circle?"

He was quite aggressive, which was in sharp contrast to how other examiners treated students. Besides, the other examiners never asked anything before the student had a chance to fully present their answer to the question on the ticket.

I said, "A circle is the set of points on the plane equidistant from a given point."

This was the standard definition.

"Wrong!" declared the man cheerfully.

How could this possibly be wrong? He waited for a few seconds and then said, "It's the set of *all* points on the plane equidistant from a given point."

That sounded like excessive parsing of words – the first sign of trouble ahead.

"OK," the man said, "What is the definition of a triangle?"

After I gave that definition, and he thought about it, no doubt trying to see if he could do some more nit-picking, he continued: "And what's the definition of a circle inscribed in a triangle?"

That led us to the definition of the tangent line, then just "a line," and that led to other things, and soon he was asking me about Euclid's fifth postulate about the uniqueness of parallel lines, which wasn't even part of the high school program! We were talking about issues that were not even close to the question on the ticket and far beyond what I was supposed to know.

Every word I said was questioned. Every concept had to be defined, and if another concept was used in the definition, then I was immediately asked to define it as well.

Needless to say, if my last name were Ivanov, I would never be asked any of these questions. In retrospect, the prudent course of action on my part would have been to protest right away and tell the examiners that they were out of line. But it's easy to say this now. I was sixteen years old, and these men were some twenty-five years my senior. They were the officials administering an exam at Moscow State University, and I felt obligated to answer their questions as best I could.

After nearly an hour-long interrogation, we moved to the second question on my ticket. By then, other students had left, and the auditorium was empty. Apparently, I was the only student in that room who required "special care." I guess they tried to place Jewish students so that there would be no more than one or two of them in the same room.

The second question asked me to write the formula for the derivative of the ratio of two functions. I was not asked to give any definitions or proofs. The question said specifically, the formula only. But of course, the examiners insisted that I explain to them a whole chapter of the calculus book.

"What is the definition of derivative?"

The standard definition I gave involved the concept of limit.

"What is the definition of limit?" Then "What is a function?" and on and on it went again.

The question of ethnic discrimination at the entrance exams to MGU has been the subject of numerous publications. For example, in his insightful article[2] in the *Notices of the American Mathematical Society*, mathematician and educator Mark Saul used my story as an example. He aptly compared my exam to the Red Queen interrogating Alice in *Alice in Wonderland*. I knew the answers, but in this game, in which everything I said was turned against me, I couldn't possibly win.

In another article[3] on this subject in the *Notices*, journalist George G. Szpiro gave this account:

Jews – or applicants with Jewish-sounding names – were singled out at the entrance exams for special treatment.... The hurdles were raised in the oral exam. Unwanted candidates were given "killer questions" that required difficult reasoning and long computations. Some questions were impossible to solve, were stated in an ambiguous way, or had no correct answer. They were not designed to test a candidate's skill but meant to weed out "undesirables." The grueling, blatantly unfair interrogations often lasted five or six hours, even though by decree they should have been limited to three and a half. Even if a candidate's answers were correct, reasons could always be found to fail him. On one occasion a candidate was failed for answering the question "what is the definition of a circle?" with "the set of points equidistant to a given point." The correct answer, the examiner said, was "the set of all points equidistant to a given point." On another occasion an answer to the same question was deemed incorrect because the candidate had failed to stipulate that the distance had to be nonzero. When asked about the solutions to an equation, the answer "1 and 2" was declared wrong, the correct answer being, according to an examiner, "1 or 2." (On a different occasion, the

same examiner told another student the exact opposite: "1 or 2" was considered
wrong.)

But back to my exam. Another hour and a half had gone by. Then one of
the examiners said:

"OK, we are done with the questions. Here is a problem we want you to
solve."

The problem he gave me was pretty hard. The solution required the use
of the so-called Sturm principle, which was not studied in school.[4] However, I
knew about it from my correspondence courses, so I was able to solve it. As I
was working my way through the final calculations, the examiner came back.

"Are you done yet?"

"Almost."

He looked at my writings and no doubt saw that my solution was correct
and that I was just finishing my calculations.

"You know what," he said, "let me give you another problem."

Curiously, the second problem was twice as hard as the first one. I was still
able to solve it, but the examiner again interrupted me halfway through.

"Not done yet?" he said, "Try this one."

If this were a boxing match, with one of the boxers pressed in the corner,
bloodied, desperately trying to hold his own against the barrage of punches
falling on him (many of them below the belt, I might add), that would be the
equivalent of the final, deadly blow. The problem looked innocent enough at
first glance: given a circle and two points on the plane outside the circle, construct another circle passing through those two points and touching the first
circle at one point.

But the solution is in fact quite complicated. Even a professional mathematician would not necessarily be able to solve it right away. One must either
use a trick called inversion or follow an elaborate geometric construction. Neither method was studied in high school, and hence this problem should not
have been allowed on this exam.

I knew about inversion, and I realized that I could apply it here. I started
to work on the problem, but a few minutes later my interrogators came back
and sat down next to me. One of them said:

"You know, I've just talked to the deputy chairman of the admissions committee and I told him about your case. He asked me why we are still wasting
our time... Look," he pulled out an official looking form with some notes scrib-
bled on it – this was the first time I saw it. "On the first question on your ticket,

you did not give us a complete answer, you didn't even know the definition of a circle. So we have to put a minus. On the second question, your knowledge was also shaky, but OK, we give you minus plus. Then you couldn't completely solve the first problem, did not solve the second problem. And on the third? You haven't solved it either. See, we have no choice but to fail you."

I looked at my watch. More than four hours had passed by since the beginning of the exam. I was exhausted.

"Can I see my written exam?"

The other man went back to the main table and brought my exam. He put it in front of me. As I was turning the pages, I felt like I was in a surrealistic movie. All answers were correct, all solutions were correct. But there were many comments. They were all made in pencil – so that they could be easily erased, I guess – but they were all ridiculous, like someone was playing a practical joke on me. One of them still stands out in my mind: in the course of a calculation, I wrote "$\sqrt{8} > 2$." And there was a comment next to it: "not proved." Really? Other comments were no better. And what grade did they give me, for all five problems solved, with all correct answers? Not 5, not 4. It was a 3, the Russian equivalent of a C. They gave me a C for this?

I knew it was over. There was no way I could fight this system. I said, "All right."

One of the men asked, "Aren't you going to appeal?"

I knew that there was an appeal board. But what would be the point? Perhaps, I could raise my grade on the written exam from 3 to 4, but appealing the result of the oral exam would be more difficult: it would be their word against mine. And even if I could raise the grade to 3, say, then what? There were still two more exams left at which they could get me.

Here is what George Szpiro wrote in the *Notices*:[5]

> And if an applicant, against all odds, managed to pass both the written and the oral test, he or she could always be failed on the required essay on Russian literature with the set phrase "the theme has not been sufficiently elaborated." With very rare exceptions, appeals against negative decisions had no chance of success. At best they were ignored, at worst the applicant was chastised for showing "contempt for the examiners."

A bigger question was: did I really want to enroll in a university that did everything in its power to prevent me from being there? I said, "No. Actually, I'd like to withdraw my application."

Their faces lit up. No appeal meant less hassle for them, less potential for trouble.

"Sure," the talkative one said, "I'll get your stuff for you right away."

We walked out of the room and entered the elevator. The doors closed. It was just the two of us. The examiner was clearly in a good mood. He said, "You did great. A really impressive performance. I was wondering: did you go to a special math school?"

I grew up in a small town; we didn't have special math schools.

"Really? Perhaps your parents are mathematicians?"

No, they are engineers.

"Interesting... It's the first time I have seen such a strong student who did not go to a special math school."

I couldn't believe what he was saying. This man had just failed me after an unfairly administered, discriminatory, grueling, nearly five-hour long exam. For all I knew, he killed my dream of becoming a mathematician. A 16-year-old student, whose only fault was that he came from a Jewish family... And now this guy was giving me compliments and expecting me to open up to him?!

But what could I do? Yell at him, punch him in the face? I was just standing there, silent, stunned. He continued: "Let me give you advice. Go to the Moscow Institute of Oil and Gas. They have an applied mathematics program, which is quite good. They take students *like you* there."

The elevator doors opened, and a minute later he handed me my thick application folder, with a bunch of my school trophies and prizes oddly sticking out of it.

"Good luck to you," he said, but I was too exhausted to respond. My only wish was to get the hell out of there!

And then I was outside, on the giant staircase of the immense MGU building. I was breathing fresh summer air again and hearing the sounds of the big city coming from a distance. It was getting dark, and there was almost no one around. I immediately spotted my parents, who were waiting anxiously for me on the steps this whole time. By the look on my face, and the big folder I was holding in my hands, they knew right away what had happened inside.

Chapter 4

Kerosinka

That night, after the exam, my parents and I came home quite late. We were still in the state of initial shock and disbelief about what happened.

This was a gut-wrenching experience for both of my parents. I have always been very close to them, and they always gave me unconditional love and support. They never pushed me to study harder or choose a particular profession, but they encouraged me to pursue my passion. And of course they were proud of my accomplishments. They were devastated by what had happened at my exam, both because of the sheer unfairness of it and because they were unable to do anything to protect their son.

Thirty years earlier, in 1954, my father's dream of becoming a theoretical physicist had been shattered just as ruthlessly, for a different reason. Like millions of innocent people, his father, my grandfather, had been a victim of Joseph Stalin's persecution. He was arrested in 1948 on bogus charges that he wanted to blow up the big automobile plant in Gorky (now Nizhny Novgorod), where he worked as the head of supplies. The only "evidence" presented in his indictment was that he had in his possession at the time of his arrest a box of matches. He was sent to a hard-labor camp at a coal mine in the northern part of Russia, part of the Gulag Archipelago that Alexander Solzhenitsyn and other writers described so vividly years later. He was deemed an "enemy of the people," and my father was therefore a "son of the enemy of the people."

My dad was obligated to write this on his application to the physics department of Gorky University. Even though he finished high school with honors and was supposed to be accepted automatically, he was failed at the interview, whose sole purpose was to screen out the relatives of the "enemies of the people." My dad was forced to go to an engineering school instead. (Like other prisoners, his father was rehabilitated and released by Nikita Khrushchev's decree in 1956, but by then it was too late to undo the injustice.)

Now, thirty years later, his son had to go through a similar experience.

But there was no time for self-pity. We had to decide quickly what to do next, and the first question was which school I should apply to. All of them held their exams at the same time, in August, about two weeks later, and I could only apply to one.

The next morning my dad woke up early and went back to Moscow. He took the recommendation from my examiner at MGU seriously. It sounded like the examiner was trying to help me, perhaps, as some partial compensation for the injustice he had done. So when my father arrived in Moscow, he went straight to the admissions office at the Institute of Oil and Gas.* Somehow my dad managed to find someone there willing to talk to him privately and described my situation. The fellow said that he was aware of the anti-Semitism at MGU but said the Institute of Oil and Gas had none of this. He went on to say that the level of applicants to their applied mathematics program was quite high due to a large number of students like me, who were not accepted at MGU. The entrance exam would be no cakewalk. But, he said, "if your son is as bright as you say he is, he will be admitted. There is no discrimination against Jews at the entrance exams here."

"I have to warn you, though," he said at the end of the conversation, "Our post-graduate studies are handled by different people, and I think your son probably won't be accepted to the grad school."

But that was something to worry about in five years, too far ahead.

My father went to a couple of other schools in Moscow with applied math programs, but there was nothing like the attitude he found at the Institute of Oil and Gas. So when he came back home that evening and told me and my mom the news, we decided right away that I would apply to the Institute of Oil and Gas, to their applied mathematics program.

The Institute was one of a dozen schools in Moscow preparing technicians for various industries, such as the Institute of Metallurgy and the Institute of Railway Engineers (in the Soviet Union, many colleges were called "institutes"). From the late 1960s, anti-Semitism at MGU "created a market for placements in mathematics for Jewish students," writes Mark Saul in his arti-

---

*At the time, it was known as the Gubkin Institute of Petrochemical and Gas Industry (it was named after the long-time head of the Ministry of Oil and Gas in the USSR, I.M. Gubkin). After I became a student there, it was renamed Gubkin Institute of Oil and Gas, and later, Gubkin University of Oil and Gas.

cle.[1] The Institute of Oil and Gas "began to cater to these markets, benefiting from the anti-Semitic policies of other universities to get highly qualified students." Mark Saul explains:

> Its nickname, Kerosinka, reflected [their] pride and cynicism. A kerosinka is a kerosene-burning space heater, a low-tech but effective response to adversity. The students and graduates of the institute quickly became known as "kerosineshchiks," and the school became a haven for Jewish students with a passion for mathematics.

> How did fate choose Kerosinka as the repository of so much talent? This question is not easy to answer. We know that there were other institutions that benefited from the exclusion of Jews from MGU. We also know that the establishment of this exclusionary policy was a conscious act, which probably met with some resistance at first. It may have been easier for some institutions to continue accepting Jewish students than for them to institute a new policy. But once the phenomenon grew and there was a cadre of Jewish students at Kerosinka, why was it tolerated? There are dark whispers of a plot by the secret police (KGB) to keep the Jewish students under surveillance in one or two places. But some of the motivation may have been more positive: the administration of the institute may have seen a good department developing and done what it could to preserve the phenomenon.

I believe the last sentence is more accurate. The President (or Rector, as he was called) of the Institute of Oil and Gas, Vladimir Nikolaevich Vinogradov, was a clever administrator known for recruiting professors who were engaged in innovative teaching and research and for using new technologies in the classrooms. He instituted the policy that all exams (including the entrance exams) were given *in writing*. Of course, there might still be some opportunity for abuse even with written tests (as was the case with my written exam at MGU), but the policy would prevent the kind of debacle that happened at my oral examination at MGU. I would not be surprised if it was Vinogradov's personal decision not to discriminate against Jewish applicants, and if so, it must have required some good will, and perhaps even some courage, on his part.

As predicted, there seemed to be no discrimination at the entrance exams. I was accepted after the first exam (written math), on which I got a 5, that is, an A (gold medalists were accepted outright if they got an A at the first exam). In a bizarre twist, this 5 did not come easy to me because apparently some of my solutions were entered incorrectly into the automated grading system, and as a result my grade was initially recorded as 4, or B. I had to go through the appeals process, which meant waiting in line for hours, with all kinds of bad thoughts swirling in my head. But once I got in to speak with the appeals committee, the error was found and fixed swiftly, an apology was offered, and my entrance exams saga came to a close.

On September 1, 1984, the school year began, and I met my new classmates. Only fifty students were accepted every year to this program (in contrast, at *Mekh-Mat* the number was close to 500). Many of my fellow students went through the same experience as I did. These were some of the brightest, most talented math students around.

Everybody, except for me and another student, Misha Smolyak from Kishinev who became my roommate at the dorm, were from Moscow. Those who lived outside of Moscow could apply only if they had graduated from high school with a gold medal, which fortunately I had.

Many of my fellow students had studied at the best Moscow schools with special math programs: schools No. 57, No. 179, No. 91, and No. 2. Some of them went on to become professional mathematicians and now work as professors at some of the best universities in the world. Just in my class, we had some of the best mathematicians of our generation: Pasha Etingof, now professor at MIT; Dima Kleinbock, professor at Brandeis University; and Misha Finkelberg, professor at the Higher School of Economics in Moscow. It was a very stimulating environment.

Mathematics was taught at Kerosinka at a high level, and basic courses, such as analysis, functional analysis, and linear algebra, were taught at the same level of rigor as at MGU. But courses in other areas of pure math, such as geometry and topology, were not available. Kerosinka only offered the applied mathematics program, so our education was geared toward concrete applications, particularly, to oil and gas exploration and production. We had to take quite a few courses of more applied orientation: optimization, numerical analysis, probability, and statistics. There was also a large computer science component.

I was glad that I had the opportunity to be exposed to these applied math courses. This taught me that there isn't really a sharp distinction between "pure" and "applied" math; good-quality applied math is always based on sophisticated pure math. But, however useful this experience was, I could not forget my true love. I knew I had to find a way to learn the pure math subjects that were not offered at Kerosinka.

The solution presented itself as I became friends with the other students, including those who went to the prestigious special math schools in Moscow. We exchanged our stories. Those who were Jewish (according to the standards I described earlier) were also failed at the exams, as ruthlessly as I was, while all of their classmates who were not Jewish were accepted to MGU without any problems. Through these other students, they knew what was happening at the *Mekh-Mat*, which courses were good, and where and when the lectures were held. So my second week at Kerosinka, my classmate (I think it was Dima Kleinbock) came up to me: "Hey, we are going to Kirillov's course at MGU. Wanna come with us?"

Kirillov was a famous mathematician, and of course I wanted to attend his lectures. But I had no idea how this would be possible. The grand building of MGU was heavily guarded by police. One needed to have a special ID to get in.

"No worries," my classmate said, "we'll scale the fence."

That sounded dangerous and exciting, so I said, "Sure."

The fence on the side of the building was quite high, easily twenty feet, but at one point the metal was bent, and it was possible to sneak in. Then what? We entered the building through a side door and after following some long corridors ended up in the kitchen. From there, through the kitchen, trying not to attract too much attention of the staff working there, to the cafeteria, and then to the main entrance hall. Elevator to the fourteenth floor, where the auditorium was.

Alexander Alexandrovich Kirillov (or San Sanych, as he was affectionately called) is a charismatic lecturer and a great human being, whom I got to know quite well years later. I think he was teaching a standard undergraduate course on representation theory along the lines of his well-known book. He also had a seminar for graduate students, which we attended as well.

We got away with this thanks to Kirillov's good heart. His son Shurik (now professor at the Stony Brook University) studied at the special math school No. 179 together with my classmates Dima Kleinbock and Syoma Hawkin. Needless to say, San Sanych knew about the situation with admissions at MGU. He told me many years later that there was nothing he could do about this – they wouldn't let him anywhere near the admissions committee, which was largely staffed with the Communist Party apparatchiks. So all he could do was let us sneak into his classes.

Kirillov did all he could to make Kerosinka students coming to his lectures feel welcome. One of the best memories of my first college year was coming to his lively lectures and seminars. I also attended a seminar given by Alexander

Rudakov, which was also a great experience.

In the meantime, I was learning whatever math I could learn at Kerosinka. I was living in a dorm but coming home for the weekends, and I was still meeting Evgeny Evgenievich every couple of weeks. He advised me on what books to read, and I reported to him on my progress. But I was quickly reaching the point where to maintain my momentum, as well as the motivation for it, I would need an advisor with whom I would meet more regularly and not only learn from, but also get a problem to work on. Because I was not at *Mekh-Mat*, I could not take advantage of the vast resources that it had to offer. And I was too shy to come up to someone like A.A. Kirillov and ask him to work with me individually, or give me a problem to work on. I felt like an outsider. By the spring semester of 1986 (my second year at Kerosinka), complacency and stagnation were beginning to set in. With all the odds stacked against me, I started to doubt that I could fulfill my dream of becoming a mathematician.

Chapter 5

Threads of the Solution

I was beginning to despair when, one day, during a break in the lecture at Kerosinka, one of our most respected math professors, Alexander Nikolaevich Varchenko, approached me in the corridor. Varchenko is a former student of Vladimir Arnold, one of the leading Soviet mathematicians, and he is a world-class mathematician himself.

"Would you be interested in working on a math problem?" he asked.

"Yes, of course," I said, "What kind of problem?" as if I would not have been happy to do just anything.

"There is this question that came up in my research, which I think is a good problem to give to a bright student like you. The expert on this matter is Dmitry Borisovich Fuchs." That was the name of a famous mathematician, which I had heard before. "I have already spoken to him, and he has agreed to supervise a student's research on this topic. Here is his phone number. Give him a call, and he'll tell you what to do."

It is quite common for experienced mathematicians like Varchenko to encounter all kinds of unsolved mathematical problems in their research. If Varchenko's problem had been closely tied to his own research program, he might have tried to solve it himself. But no mathematician does everything alone, so mathematicians often delegate some of such unsolved problems (typically, the ones they consider to be simpler) to their students. Sometimes a problem might be outside of the professor's immediate interests, but he or she might nonetheless be curious about it, as was the case with my problem. That's why Varchenko enlisted Fuchs, an expert in this area, to supervise me. All in all, this was for the most part a typical "transaction" in the social workings of the mathematical world.

What was actually unusual was that Fuchs was not formally affiliated with teaching at any university. But for many years Fuchs had been, along with a number of other top mathematicians, trying to alleviate the effect of the discrimination against Jewish students by privately teaching young talented kids who were denied entry to MGU.

As part of those efforts, Fuchs was involved in what became known as "Jewish People's University," an unofficial evening school, where he and his colleagues gave courses of lectures to students. Some of those lectures had even been held at Kerosinka, although this was before my time.

The school had been organized by a courageous woman, Bella Muchnik Subbotovskaya, who was its heart and soul. Unfortunately, the KGB got on the case, alarmed that there were unauthorized gatherings of Jewish people. She was eventually called to the KGB and interrogated. Soon after that interview, she was killed by a truck under suspicious circumstances, which led many people to suspect that this was in fact a cold-blooded murder.[1] Without her at the helm, the school collapsed.

I came to Kerosinka two years after this tragic chain of events. Though the evening school did not exist anymore, there was still a small network of professional mathematicians who helped misfortunate outcasts like myself on an individual basis. They sought out promising students and gave them advice, encouragement, and in some cases, full-fledged mentoring and advising. This was the reason that Varchenko gave that problem to me, a student at Kerosinka, rather than a student at *Mekh-Mat*, where, through his connections, he could have easily found a student willing to take it up. This was also why Fuchs was willing to invest his personal time to supervise me.

And I am glad he did. Looking back, it is clear to me that without Fuchs' kindness and generosity, I would have never become a mathematician. I was studying math at Kerosinka and sitting in at the lectures at MGU, but by itself that was not enough. In fact, it is virtually impossible for students to do their own research without someone guiding their work. Having an advisor is absolutely essential.

At the time, though, all I knew was that I had in my hand the phone number of Fuchs, a renowned mathematician, and I was about to embark on a project supervised by him. This was unbelievable! I didn't know where this would end up, but I knew right away that something big had happened.

That evening, having mustered all my courage, I called Fuchs from a pay phone and explained who I was.

"Yes, I know," said Fuchs, "I have to give you a paper to read."

We met the next day. Fuchs had the physical appearance of a giant, not at all how I imagined him. He was very business-like.

"Here," he said, handing me an offprint of an article, "try to read this, and as soon as you see a word that you don't understand, call me."

I felt like he had just handed me the Holy Grail.

This was an article, a dozen pages long, which he had written some years earlier, on the subject of "braid groups." That evening I started reading it.

The preceding three years of studying with Evgeny Evgenievich and on my own were not spent in vain. Not only did I understand all the words in the title, I could make sense of the content as well. I decided to try to read the whole thing on my own. It was a matter of pride. I was already imagining how impressed Fuchs would be when I told him that I understood everything on my own.

I had heard of the "braid groups" before. These are excellent examples of groups, the concept we discussed in Chapter 2. Evgeny Evgenievich had introduced this concept in the context of symmetries, and so elements of the groups that we considered were symmetries of some object. For example, the circle group consisted of the symmetries of a round table (or any other round object), and the group of four rotations was the group of symmetries of a square table (or any other square object). Once we have the notion "group," we can look for other examples. It turns out there are many examples of groups that have nothing to do with symmetries, which was our motivation to introduce the concept of a group in the first place. This is actually a typical story. The creation of a mathematical concept may be motivated by problems and phenomena in one area of math (or physics, engineering, and so forth), but later it may well turn out to be useful and well adapted to other areas.

It turns out that many groups do *not* come from symmetries. And the braid groups are such groups.

I did not know yet about the real-world applications of braid groups to such areas as cryptography, quantum computing, and biology, which we will talk about later. But I was mesmerized by the innate beauty of these mathematical abstractions.

There is one braid group for each natural number $n = 1, 2, 3,...$ We can use those numbers to get a name for each braid group. In general, we call them B_n, and so for $n = 1$ we have a group called B_1, for $n = 2$ we have a group called B_2, and so on.

To describe the group B_n, we have to describe first its elements, as we did with the rotational symmetries of the round and square tables. The elements of the group B_n are the so-called *braids with n threads*, such as the one shown on the picture below, with $n = 5$. Imagine two solid, transparent plates with five nails in each, with one thread connecting each nail in one plate to one nail in the other. Since the plates are transparent, we can see each of the threads in its entirety. Each thread is allowed to weave around any other thread any way we like but is *not* allowed to get entangled with itself. Each nail must connect to exactly one thread. The positions of the plates are fixed once and for all.

This whole thing – two plates and however many threads – constitutes a single braid, just as a car has four wheels, one transmission, four doors, and so forth. We are not considering those parts separately; we are focusing on the braid as a whole.

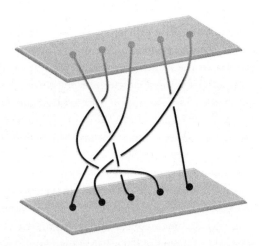

Those are the braids with n threads. Now we need to show that all braids with n threads form a group. This means that we need to describe how to make the composition of two such braids. In other words, for each pair of braids with n threads, we have to produce another braid with n threads, just as applying two rotations one after another gave us a third rotation. And then we will have to check that this composition satisfies the properties listed in Chapter 2.

So suppose we have two braids. In order to produce a new braid out of them, we put one of them on top of the other, aligning the nails, as shown on the picture. And then we remove the middle plates while connecting the upper threads to the lower ones attached to the matching nails.

The resulting braid will be twice as tall, but this is not a problem. We'll just shorten the threads so that the resulting braid will have the same height as the original ones, while preserving the way the threads go around each other. *Voilà!* We started out with two braids and produced a new one. This is the rule of composition of two braids in our braid group.

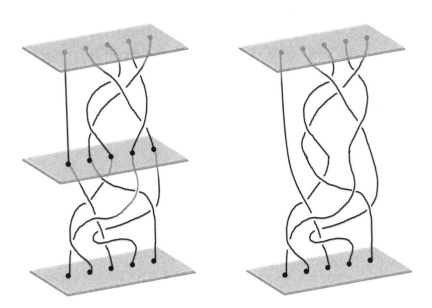

Since a braid group does not come from symmetries, it is sometimes better to think of this operation not as "composition" (which was natural in the case of groups of symmetries), but as "addition" or "multiplication," similar to the operations that we perform on numbers. From this point of view, braids are like numbers – these are some "hairy numbers," if you will.

Given two whole numbers, we can add them to each other and produce a new number. Likewise, given two braids, we produce a new one by the rule described above. So we can think of this as the "addition" of two braids.

Now we need to check that this addition of braids satisfies all properties (or axioms) of a group. First, we need the identity element. (In the circle group, this was the point corresponding to the rotation by 0 degrees.) This will be the braid with all threads going straight down without any weaving as shown on the next picture. It is a kind of "trivial" braid, in which no braiding actually occurs, the same way rotation by 0 degrees makes no rotation at all.[2]

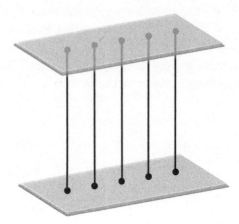

Next, we need to find the inverse braid of a given braid b (in the case of the circle group, this was the rotation by the same angle but in the opposite direction). It should be such that if we add this braid to the braid b, according to the rule described above, we will get the identity braid.

This inverse braid will be the reflection of b with respect to the bottom plate. If we compose it with the original braid according to our rule, we will be able to rearrange all threads so that the result will be the identity braid.

Here I need to make an important point, which up to now I have kind of swept under the rug: we will not distinguish the braids that can be obtained from one another by pulling the threads, stretching and shrinking them any way we like so long as we do not cut or resew the threads. In other words, the threads should be attached to the same nails, and we do not allow the threads to go through each other, but otherwise we can tweak them any way we like. Think of this as grooming our braid. When we do that, it will still be the same braid (only prettier!). It is in this sense that the addition of a braid and its mirror image is "the same" as the identity braid; it is not literally the same but becomes one after we tweak the threads.[3]

So we see now that the axioms of a group – composition (or addition), identity, and inverse – are satisfied. We have proved that braids with n threads form a group.[4]

To see what the braid groups are more concretely, let's look closely at the simplest one: the group B_2 of braids with two threads. (The group with B_1 with one thread has only one element, so there is nothing to discuss.[5]) We will assign to each such braid an integer N. By an integer, here I mean a natural number: 1, 2, 3,...; or 0; or a negative of a natural number: -1, -2, -3,...

First of all, to the identity braid we will assign the number 0. Second, if the thread starting at the left nail on the top plate goes underneath the other thread, then we assign to it 1. If it goes around it, then we assign to it 2, and so on, as shown on the pictures.

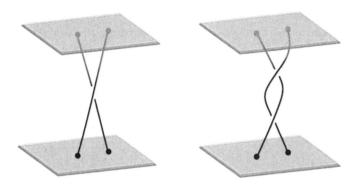

If this thread goes on top of the other thread, then we assign to the braid the negative number -1; if it goes around it as shown on the picture below, we assign to it -2; and so on.

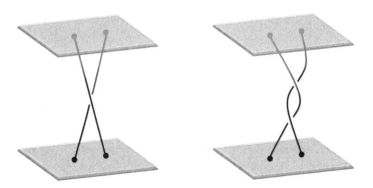

Let's call the number assigned to a braid in this way the "number of over-laps." If we have two braids with the same number of overlaps, we can transform one into another by "tweaking" the threads. In other words, the braid is completely determined by the number of overlaps. So we have a one-to-one correspondence between braids with two threads and integers.

Here it is useful to note something we always take for granted: the set of all integers is itself a group! Namely, we have the operation of addition, the "identity element" is number 0, and for any integer N, its "inverse" is $-N$. Then all properties of a group listed in Chapter 2 are satisfied. Indeed, we have $N + 0 = N$ and $N + (-N) = 0$.

What we have just found is that the group of braids with two threads has the same structure as the group of integers.[6]

Now, in the group of integers the sum of two integers a and b is the same in two different orders:

$$a + b = b + a.$$

This is also so in the braid group B_2. Groups satisfying this property are called "commutative" or "abelian" (in honor of the Norwegian mathematician Niels Henrik Abel).

In a braid with 3 threads or more, the threads can be entangled among themselves in a much more complicated fashion than in a braid with only 2 threads. The knotting pattern can no longer be described merely by the numbers of overlaps (look at the above picture of a braid with 5 threads). The pattern in which the overlaps occur is also important. Furthermore, it turns out that the addition of two braids with 3 or more threads does depend on the order in which it is taken (that is to say, which of the two braids is on top in the picture above, describing the addition of braids). In other words, in the group B_n with $n = 3, 4, 5, \ldots$ we have in general

$$a + b \neq b + a.$$

Such groups are called "non-commutative" or "non-abelian."

Braid groups have many important practical applications. For example, they are used to construct efficient and robust public key encryption algorithms.[7]

Another promising direction is designing quantum computers based on creating complex braids of quantum particles known as anyons. Their trajectories weave around each other, and their overlaps are used to build "logic gates" of the quantum computer.[8]

There are also applications in biology. Given a braid with n threads, we can number the nails on the two plates from 1 to n from left to right. Then, connect the ends of the threads attached to the nails with the same number on the two plates. This will create what mathematicians call a "link": a union of loops weaving around each other.

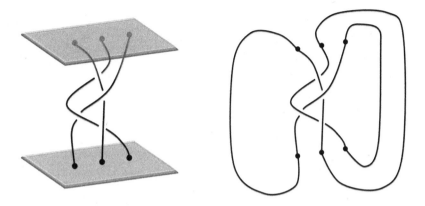

In the example shown on this picture, there is only one loop. Mathematicians' name for it is "knot." In general, there will be several closed threads.

The mathematical theory of links and knots is used in biology: for example, to study bindings of DNA and enzymes.[9] We view a DNA molecule as one thread, and the enzyme molecule as another thread. It turns out that when they bind together, highly non-trivial knotting between them may occur, which may alter the DNA. The way they entangle is therefore of great importance. It turns out that the mathematical study of the resulting links sheds new light on the mechanisms of recombination of DNA.

In mathematics, braids are also important because of their geometric interpretation. To explain it, consider all possible collections of n points on the plane. We will assume that the points are distinct; that is, for any two points, their positions on the plane must be different. Let's choose one such collection; namely, n points arranged on a straight line, with the same distance between neighboring points. Think of each point as a little bug. As we turn on the music, these bugs come alive and start moving on the plane. If we view the time as the vertical direction, then the trajectory of each bug will look like a thread. If the positions of the bugs on the plane are distinct at all times – that is, if we assume that the bugs don't collide – then these threads will never intersect. While the music is playing, they can move around each other, just like the threads of a braid. However, we demand that when we stop the music after a fixed period of time, the bugs must align on a straight line in the same way as at the beginning, but each bug is allowed to end up in a position initially occupied by another bug. Then their collective path will look like a braid with n threads.

Thus, braids with n threads may be viewed as paths in the space of collections of n distinct points on the plane.[10]

The problem that Varchenko gave me, and on which I was about to start working with Fuchs, concerned a part of the braid group called the "commutator subgroup." Remember that for braids with two threads, we have defined the overlap number. A similar number can be assigned to a braid with any number of threads.[11] We use this to define the commutator subgroup B'_n of the braid group with n threads. It consists of all braids whose total overlap number is zero.[12]

The problem I had to solve was to compute the so-called "Betti numbers" of the group B'_n. These numbers reflect deep properties of this group, which are important in applications. As an analogy, think of a physical object, like a house. It has various characteristics: some more obvious, like the numbers of floors, rooms, doors, windows, etc., and some less so, like the proportions of the materials from which it is built. Likewise, a group also has various characteristics, and these are the Betti numbers.[13] Fuchs had earlier computed the Betti numbers of the braid group B_n itself. He gave me his paper so that I could learn the basics of the subject.

Within a week, I was able to read the entire Fuchs paper on my own, occasionally looking up previously unknown-to-me concepts and definitions in my by-then fairly large library of math books. I called Fuchs.

"Oh, it's you," he said. "I was wondering why you hadn't called. Have you started reading the article?"

"Yes, Dmitry Borisovich. Actually, I have finished it."

"Finished it?" Fuchs sounded surprised. "Well, then we should meet. I want to hear what you've learned."

Fuchs suggested we meet on the next day at MGU, after a seminar he was going to attend. As I was preparing for the meeting, I kept re-reading the article and practicing my answers to the kinds of questions that I thought Fuchs was likely to ask. A world-class mathematician like Fuchs wouldn't just take up a new student out of pity. The bar was set high. I understood that my first conversation with Fuchs would be something of an audition, and that's why I was so eager to make a good impression on him.

We met at the appointed hour and walked the corridors of *Mekh-Mat* to find a bench where we wouldn't be bothered. After we sat down, I started telling Fuchs what I learned from his article. He listened attentively, occasionally asking me questions. I think he was pleased by what he was hearing. He was curious where I learned all this stuff, and I told him about my studies with Evgeny Evgenievich, reading books, and attending lectures at *Mekh-Mat*. We even talked about my exam at the MGU (this was of course nothing new to Fuchs).

Luckily, our meeting went well. Fuchs seemed impressed with my knowledge. He told me that I was ready to tackle Varchenko's problem and that he would help me with it.

I was elated when I was leaving MGU that evening. I was about to start working on my first math problem, guided by one of the best mathematicians in the world. Less than two years had passed since my entrance exam at *Mekh-Mat*. I was back in the game.

Chapter 6

Apprentice Mathematician

Solving a mathematical problem is like doing a jigsaw puzzle, except you don't know in advance what the final picture will look like. It could be hard, it could be easy, or it could be impossible to solve. You never know until you actually do it (or realize that it's impossible to do). This uncertainty is perhaps the most difficult aspect of being a mathematician. In other disciplines, you can improvise, come up with different solutions, even change the rules of the game. Even the very notion of what constitutes a solution is not clearly defined. For example, if we are tasked with improving productivity in a company, what metrics do we use to measure success? Will an improvement by 20 percent count as a solution of the problem? How about 10 percent? In math, the problem is always well defined, and there is no ambiguity about what solving it means. You either solve it or you don't.

For Fuchs' problem, I had to compute the Betti numbers of the groups B'_n. There was no ambiguity in what this meant. It means the same thing today to everyone familiar with the language of math as it did in 1986 when I first learned about this problem, and will mean the same thing a hundred years from now.

I knew that Fuchs had solved a similar problem, and I knew how he'd done it. I prepared for my own task by working on similar problems for which solutions had already been known. This gave me intuition, skills, and a toolkit of methods. But I could not know *a priori* which of these methods would work or which way I should approach the problem – or even whether I could solve it without creating an essentially new technique or an entirely different method.

This quandary besets all mathematicians. Let's look at one of the most famous problems in mathematics, Fermat's Last Theorem, to see how one can go about doing math when the problem is easy to state but the solution is far from obvious. Fix a natural number n, that is, 1, 2, 3,..., and consider the equation

$$x^n + y^n = z^n$$

on the natural numbers $x, y,$ and z.

If $n = 1$, we get the equation

$$x + y = z,$$

which surely has many solutions among natural numbers: just take any x and y and set $z = x + y$. Note that here we use the operation of addition of natural numbers that we discussed in the previous chapter.

If $n = 2$, we get the equation

$$x^2 + y^2 = z^2.$$

This equation also has many solutions in natural numbers; for instance,

$$3^2 + 4^2 = 5^2.$$

All of this has been known since antiquity. What was unknown was whether the equation had any solutions for n greater than 2. Sounds pretty simple, right? How hard could it be to answer a question like this?

Well, as it turned out, pretty hard. In 1637, a French mathematician, Pierre Fermat, left a note on the margin of an old book saying that if n is greater than 2, then the equation had no solutions x, y, z that are natural numbers. In other words, we cannot find three natural numbers x, y, z such that

$$x^3 + y^3 = z^3,$$

cannot find natural numbers x, y, z such that

$$x^4 + y^4 = z^4,$$

and so on.

Fermat wrote that he had found a simple proof of this statement, for all n greater than 2, but "this margin is too small to contain it." Many people, professional mathematicians as well as amateurs, took Fermat's note as a challenge and tried to reproduce his "proof," making this the most famous mathematical problem of all time. Prizes were announced. Hundreds of proofs were written and published, only to be crashed later on. The problem remained unsolved 350 years later.

In 1993, a Princeton mathematician, Andrew Wiles, announced his own proof of Fermat's Last Theorem. But his proof, at first glance, had nothing to do with the original problem. Instead of proving Fermat's Last Theorem, Wiles tackled the so-called Shimura–Taniyama–Weil conjecture, which is about something entirely different and is a lot more complicated to state. But a few years earlier, a Berkeley mathematician named Ken Ribet had proved that the statement of this conjecture implies Fermat's Last Theorem. That's why a proof of the conjecture would also prove Fermat's Last Theorem. We will talk about all this in detail in Chapter 8; the point I want to make now is that what looks like a simple problem may not necessarily have an elementary solution. It is clear to us now that Fermat could not have possibly proved the statement attributed to him. Entire fields of mathematics had to be created in order to do this, a development that took a lot of hard work by many generations of mathematicians.[1]

But is it possible to predict all that, given this innocent-looking equation?

$$x^n + y^n = z^n$$

Not at all!

With any math problem, you never know what the solution will involve. You hope and pray that you will be able to find a nice and elegant solution, and perhaps discover something interesting along the way. And you certainly hope that you will actually be able to do it in a reasonable period of time, that you won't have to wait for 350 years to reach the conclusion. But you can never be sure.

In the case of my problem, I was lucky; there was in fact an elegant solution that I was able to find in a relatively short period of time, about two months. But it didn't come easily to me. It never does. I tried many different methods. As each of them failed, I felt increasingly frustrated and anxious. This was my first problem, and inevitably I questioned whether I could be a mathematician. This problem was my first test of whether I had what it takes.

Working on this problem didn't excuse me from taking classes and passing exams at Kerosinka, but my highest priority was the problem, and I spent endless hours with it, nights and weekends. I was putting way too much pressure on myself. I was starting to have trouble sleeping, the first time this ever happened to me. The insomnia I acquired while working on this problem was the first "side effect" of my mathematical research. It haunted me for many months afterward, and from that point on I never allowed myself to get lost so completely in a math problem.

I met with Fuchs every week or so at the *Mekh-Mat* building, where I told him about my progress, or lack of it (by then he was able to get me an ID, so I did not have to scale the fence anymore). Fuchs was always supportive and encouraging, and each time we met he would tell me about a new trick or suggest a new insight, which I would try to apply to my problem.

And then, suddenly, I had it. I found the solution, or perhaps more accurately, the solution presented itself, in all of its splendor.

I was trying to use one of the standard methods for computing Betti numbers, which Fuchs had taught me, called "spectral sequence." I was able to apply it in a certain way, which allowed me in principle to compute the Betti numbers of the group B'_n from the knowledge of the Betti numbers of all the groups B'_m with $m < n$. The caveat was, of course, that I did not know what those other Betti numbers were either.

But this gave me a way to attack the problem: if I could *guess* the right answer, I would then have a path to *proving* it by following this method.

That's easy to say, but coming up with such a guess required many sample computations, which only became more and more complicated. For a long time, no pattern seemed to emerge.

Suddenly, as if in a stroke of black magic, it all became clear to me. The jigsaw puzzle was complete, and the final image was revealed to me, full of elegance and beauty, in a moment that I will always remember and cherish. It was an incredible feeling of high that made all those sleepless nights worthwhile.

For the first time in my life, I had in my possession something that *no one else in the world* had. I was able to say something new about the universe. It wasn't a cure for cancer, but it was a worthy piece of knowledge, and no one could ever take it away from me.

If you experience this feeling once, you will want to go back and do it again. This was the first time it happened to me, and like the first kiss, it was very special. I knew then that I could call myself mathematician.

The answer was actually quite unexpected, and much more interesting than what Fuchs or I could imagine. I found that for each divisor of the natural number n (the number of threads in the braids we are considering), there is a Betti number of the group B'_n that is equal to the celebrated "Euler function" of that divisor.[2]

The Euler function assigns to any natural number d another natural number, called $\phi(d)$. This is the number of integers between 1 and d that are *relatively prime* with d; that is, have no common divisors with d (apart from 1, of course).

For example, take $d = 6$. Then 1 is relatively prime with 6, 2 is not (it is a divisor of 6), 3 is not (it is also a divisor of 6), 4 is not (4 and 6 share a common divisor; namely, 2), 5 is relatively prime with 6, and 6 is not. So there are two natural numbers between 1 and 6 that are relatively prime with 6: namely, 1 and 5. Hence the Euler function of 6 is equal to 2. We write this as $\phi(6) = 2$.

The Euler function has many applications. For example, it is employed in the so-called RSA algorithm used to encrypt credit card numbers in online transactions (this is explained in endnote 7 to Chapter 14). It is named in honor of the eighteenth-century Swiss mathematician Leonhard Euler.

The fact that the Betti numbers I found were given by the Euler function suggested the existence of some hidden connections between braid groups and number theory. Therefore, the problem I had solved could potentially have implications far beyond its original scope.

Of course, I was eager to tell Fuchs about my results. It was already June 1986, almost three months after he and I first met. By then, Fuchs had left Moscow with his wife and two young daughters to spend the summer at his dacha near Moscow. Luckily for me, it was situated along the same train line as my hometown, about halfway, and so it was easy for me to visit him there on my way home.

After offering me a customary cup of tea, Fuchs asked me about my progress.

"I solved the problem!"

I couldn't contain my excitement, and I guess the account of the proof that I gave was pretty rambling. But no worries – Fuchs understood everything quickly. He looked pleased.

"This is great," he said, "Well done! Now you have to start writing a paper about this."

It was the first time I wrote a math paper, and it turned out to be no less frustrating than my mathematical work, but much less fun. Searching for new patterns on the edge of knowledge was captivating and exciting. Sitting at my desk, trying to organize my thoughts and put them on paper, was an entirely different process. As someone told me later, writing papers was the punishment we had to endure for the thrill of discovering new mathematics. This was the first time I was so punished.

I came back to Fuchs with different drafts, and he read them carefully, pointing out deficiencies and suggesting improvements. As always, he was extremely generous with his help. From the beginning, I put Fuchs' name as one of the coauthors, but he flatly refused. "This is your paper," he said. Finally, Fuchs declared that the article was ready, and he told me that I should submit it to *Functional Analysis and Applications,* the math journal run by Israel Moiseevich Gelfand, the patriarch of the Soviet mathematical school.

A compact charismatic man, then in his early seventies, Gelfand was a legend in the Moscow mathematical community. He presided over a weekly seminar held at a grand auditorium on the fourteenth floor of the main MGU building. This was an important mathematical and social event, which had been running for more than fifty years and was renowned all over the world. Fuchs was a former collaborator of Gelfand (their work on what became known as "Gelfand–Fuchs cohomology" was widely known and appreciated) and one of the most senior members of Gelfand's seminar. (The others included A.A. Kirillov, who was Gelfand's former student, and M.I. Graev, Gelfand's longtime collaborator).

The seminar was unlike any other seminars I have ever attended. Usually, a seminar has fixed hours – in the U.S. one hour or an hour and a half – and there is a speaker who prepares a talk on a particular topic chosen in advance. Occasionally, the audience members ask questions. It was not at all like this at the Gelfand seminar. It met every Monday evening, and the official starting time was 7:00 pm. However, the seminar rarely started before 7:30, and it usually began around 7:45 to 8:00. During the hour or so before the start, the members of the seminar, including Gelfand himself (who usually arrived around 7:15–7:30), would wander around and talk to each other inside the auditorium and in the large foyer outside. Clearly, this was what Gelfand had intended. This was as much a social event as a math seminar.

Most of the mathematicians coming to the Gelfand seminar worked at various places that were not affiliated with MGU. Gelfand's seminar was the only place where they could meet their peers, find out what was happening in the world of mathematics, share their ideas, and forge collaborations. Since Gelfand was himself Jewish, his seminar was considered as one of "safe havens" for Jews and even hailed as "the only game in town" (or one of very few) in which

Jewish mathematicians could participate (though, in fairness, many other seminars at MGU were open to the public and were run by people who were not prejudiced against any ethnicities). No doubt, Gelfand gladly took advantage of this.

The anti-Semitism that I had experienced at my entrance exam to MGU spread to all levels of academia in the Soviet Union. Earlier, in the 1960s and early 1970s, even though there were restrictions, or "quotas," for students of Jewish background, they could still get in as undergraduates at the *Mekh-Mat* (the situation gradually worsened throughout the 1970s and early 1980s, to the point where in 1984, when I was applying to *Mekh-Mat*, almost no Jewish students were accepted).[3] But even in those years, it was nearly impossible for these students to enter graduate school. The only way Jewish students could do this was to go to work somewhere for three years after getting the bachelor's degree, and then one could be sent to graduate school by their employer (often, located somewhere in a faraway province). And even if they managed to overcome this hurdle and get a Ph.D., it was impossible for them to find an academic job in mathematics in Moscow (at MGU, for example). Either they had to settle for a job somewhere in the province or join one of many research institutes in Moscow that had little or nothing to do with mathematics. The situation was even more difficult for those who were not originally from Moscow, because they did not have *propiska*, a Moscow residency stamp in their interior passport, which was required for employment in the capital.

Even the most exceptional students got such treatment. Vladimir Drinfeld, a brilliant mathematician and future Fields Medal winner about whom we will talk more later, was allowed to become a graduate student at *Mekh-Mat* right after obtaining his bachelor's degree (though from what I've heard it was very difficult to arrange), but being a native of Kharkov, Ukraine, he could not be employed in Moscow. He had to settle for a teaching job at a provincial university in Ufa, an industrial city in the Ural Mountains. Eventually, he got a job as a researcher at the Institute for Low Temperature Physics in Kharkov.

Those who stayed in Moscow were employed at places like the Institute for Seismic Studies or Institute for Signal Processing. Their day jobs consisted of some tedious calculations related to a particular industry to which their institute was attached (though some actually managed to break new ground in those areas, multi-talented as they were). They had to do the kind of mathematical research that was their true passion on the side, in their spare time.

Gelfand himself was forced out of his teaching job at *Mekh-Mat* in 1968 after he signed the famous letter of ninety-nine mathematicians demanding the release of the mathematician and human rights activist Alexander Esenin-Volpin (the son of the poet Sergei Esenin) from a politically motivated detention in a psychiatric hospital. That letter was so skillfully written that after it was broadcast on the BBC radio, the worldwide outrage embarrassed the Soviet leadership to release Esenin-Volpin almost immediately.[4] But this also greatly angered the authorities. They subsequently found ways to punish everyone who signed it. In particular, many of the signatories were fired from their teaching jobs.[5]

So Gelfand was no longer professor of mathematics at MGU, though he was able to preserve his seminar that was still being held at the main MGU building. His official job was at a biological lab of MGU that he had founded to conduct research in biology, which was also his passion.[*] Fuchs was employed at the same lab.

Fuchs had earlier urged me to start attending Gelfand's seminar, and I did come to a couple of meetings at the very end of the spring semester. Those meetings made a great impression on me. Gelfand ran his seminar in the most authoritarian way. He decided its every aspect, and though to an untrained eye his seminars could appear chaotic and disorganized, he actually devoted an enormous amount of time and energy to the preparation and choreographing of the weekly meetings.

Three years later, when Gelfand asked me to speak about my work, I had the opportunity to see the inner workings of the seminar up close. For now, I was observing it from the vantage point of a seventeen-year-old student just starting his mathematical career.

The seminar was in many ways the theater of a single actor. Officially, there would be a designated speaker reporting on a designated topic, but typically only part of the seminar would be devoted to it. Gelfand would usually bring up other topics and call other mathematicians, who had not been asked to prepare in advance, to the blackboard to explain them. But he was always at the center of it all. He and only he controlled the flow of the seminar and had the absolute power to interrupt the speaker at any moment with questions, suggestions, and

---

[*]It is also worth noting that Gelfand was not elected as a full member of the Academy of Sciences of the USSR until the mid-1980s because the Mathematical Branch of the Academy was for decades controlled by the director of the Steklov Mathematical Institute in Moscow, Ivan Matveevich Vinogradov, nicknamed the "Anti-Semite-in-Chief of the USSR." Vinogradov had put in place draconian anti-Semitic policies at the Academy and the Steklov Institute, which was in his grip for almost fifty years.

remarks. I can still hear him say "*Dayte opredelenie*" – "Give the definition" – his frequent admonition to a speaker.

He also had the habit of launching into long tirades on various topics (sometimes unrelated to the material discussed), telling jokes, anecdotes, and stories of all kinds, many of them quite entertaining. This was where I heard the parable that I mentioned in the Preface: a drunkard may not know which number is larger, 2/3 or 3/5, but he knows that 2 bottles of vodka for 3 people is better than 3 bottles of vodka for 5 people. One of Gelfand's skills was his ability to "rephrase" questions asked by others in such a way that the answer became obvious.

Another joke he liked to tell involved the wireless telegraph: "At the beginning of the twentieth century, someone asks a physicist at a party: can you explain how it works? The physicist replies that it's very simple. First, you have to understand how the ordinary, wired, telegraph works: imagine a dog with its head in London and its tail in Paris. You pull the tail in Paris, and the dog barks in London. The wireless telegraph, says the physicist, is the same thing, but without a dog."

After recounting the joke and waiting for the laughter to subside (even from those people in the audience who had heard it a thousand times), Gelfand would pivot to whatever math problem was being discussed. If he thought that the solution of the problem required a radically new approach, he would comment, "What I'm trying to say is we need to do it without a dog."

A frequently used device at the seminar was to appoint a *kontrol'nyj slushatel'*, a test listener, usually a junior member of the audience, who was supposed to repeat at regular intervals what the speaker was saying. If it was deemed that the "test listener" was following the lecture well, this meant that the speaker was doing a good job. Otherwise, the speaker had to slow down and explain better. Occasionally, Gelfand would even discharge a particularly incomprehensible speaker in disgrace and replace him or her with another member of the audience. (Of course, Gelfand would poke fun at the test listener as well.) All of this made the seminar very entertaining.

Most seminars proceed at a steady pace, with people in the audience listening politely (and some perhaps dozing off) – too complacent, too polite, or simply afraid to ask the speaker any questions, and perhaps learning little. There is no doubt that the uneven pace and the generally subversive character of the Gelfand seminar not only kept people awake (not an easy task given that the seminar often lasted till midnight), but stimulated them in ways that other seminars simply couldn't. Gelfand demanded a lot of his speakers. They worked hard, and so did he. Whatever one can say about Gelfand's style, people never left the seminar empty-handed.

However, it seems to me that a seminar like this could only exist in a totalitarian society, like the Soviet Union. People were accustomed to the kind of dictatorial powers and behavior that Gelfand displayed. He could be cruel, at times insulting, to people. I don't think many would tolerate this kind of treatment in the West. But in the Soviet Union, this was not considered to be out of the ordinary, and no one protested. (Another famous example like this was Lev Landau's seminar on theoretical physics.)

When I first started coming to the seminar, Gelfand had a young physicist, Vladimir Kazakov, present a series of talks about his work on so-called matrix models. Kazakov used methods of quantum physics in a novel way to obtain deep mathematical results that mathematicians could not obtain by more conventional methods. Gelfand had always been interested in quantum physics, and this topic had traditionally played a big role at his seminar. He was particularly impressed with Kazakov's work and was actively promoting it among mathematicians. Like many of his foresights, this proved to be golden: a few years later this work became famous and fashionable, and it led to many important advances in both physics and math.

In his lectures at the seminar, Kazakov was making an admirable effort to explain his ideas to mathematicians. Gelfand was more deferential to him than usual, allowing him to speak without interruptions longer than other speakers.

While these lectures were going on, a new paper arrived, by John Harer and Don Zagier, in which they gave a beautiful solution to a very difficult combinatorial problem.[6] Zagier has a reputation for solving seemingly intractable problems; he is also very quick. The word was that the solution of this problem took him six months, and he was very proud of that. At the next seminar, as Kazakov was continuing his presentation, Gelfand asked him to solve the Harer–Zagier problem using his work on the matrix models. Gelfand had sensed that Kazakov's methods could be useful for solving this kind of problem, and he was right. Kazakov was unaware of the Harer–Zagier paper, and this was the first time he heard this question. Standing at the blackboard, he thought about it for a couple of minutes and immediately wrote down the Lagrangian of a quantum field theory that would lead to the answer using his methods.

Everyone in the audience was stunned. But not Gelfand. He asked Kazakov innocently, "Volodya, how many years have you been working on this topic?"

"I am not sure, Israel Moiseevich, perhaps six years or so."

"So it took you six years plus two minutes, and it took Don Zagier six months. Hmmm... You see how much better he is?"

And this was a mild "joke," compared with some others. You had to have thick skin in order to survive in this environment. Unfortunately, some speakers took these kinds of public put-downs personally, and this caused them a lot of torment. But I have to add that Gelfand always had a sharper tongue for the older, more established mathematicians, and he was much more gentle to young mathematicians, especially to students.

He used to say that at the seminar he welcomed all undergraduates, talented graduate students, and only brilliant professors. He understood that in order to keep the subject moving, it was very important to prepare new generations of mathematicians, and he always surrounded himself with young talent. They kept him young as well (he was actively doing cutting-edge research till he was in his late eighties). Often, he would even invite high school students to the seminar and make them sit in the front row to make sure that they were following what was going on. (Of course, these were no ordinary high school students. Many of them went on to become world-renowned mathematicians.)

By all accounts, Gelfand was very generous with his students, spending hours talking to them on a regular basis. Very few professors do this. It wasn't easy to be his student; he gave them a kind of tough love, and they had to cope with his various quirks and dictatorial habits. But my impression from talking to many of them is that they were all loyal to him and felt they owed him a tremendous debt.

I was not Gelfand's student – I was his "grand-student," as both of my teachers, Fuchs and Feigin (who was not yet in my life), were at least partially Gelfand's students. Hence I always considered myself as being part of the "Gelfand mathematical school." Much later, when he and I were in the United States, Gelfand asked me directly about this, and by the look of satisfaction on his face when I said yes, I could tell how important the issue of his school and recognition of who was part of it was to him.

This school, of which the seminar was the focal point, its window to the world, had an enormous impact not only on mathematics in Moscow, but around the world. Foreign mathematicians came to Moscow just to meet Gelfand and attend his seminar, and many considered it an honor to lecture there.

Gelfand's fascinating and larger-than-life personality played a big role in the seminar's reputation. A few years later, he became interested in my work and asked me to speak at his seminar. I spent many hours talking to him, not just about mathematics, but about a lot of other things. He was very interested in the history of mathematics and his own legacy in particular. I remember vividly how, when I first came to visit him at his Moscow apartment (I had just turned twenty-one), he informed me that he considered himself the Mozart of mathematics.

"Most composers are remembered for particular pieces they wrote," he said. "But in the case of Mozart, that's not so. It's the totality of his work that makes him a genius." He paused and continued: "The same goes for my mathematical work."

Putting aside some interesting questions raised by such self-assessment, I think it's actually an apt comparison. Though Gelfand did not prove any famous long-standing conjectures (such as Fermat's Last Theorem), the cumulative effect of his ideas on mathematics was staggering. Perhaps more importantly, Gelfand possessed an excellent taste for beautiful mathematics as well as an astute intuition about which areas of mathematics were the most interesting and promising. He was like an oracle who had the power to predict in which directions mathematics would move.

In the subject that was becoming increasingly fractured and specialized, he was one of the last remaining Renaissance men able to bridge different areas. He epitomized the unity of mathematics. Unlike most seminars, which focus on one area of math, if you came to Gelfand's seminar, you could see how all these different parts fit together. That's why all of us gathered every Monday night on the fourteenth floor of the main MGU building and eagerly awaited the word of the master.

And it was to this awe-inspiring man that Fuchs suggested I submit my first math paper. Gelfand's journal, *Functional Analysis and Applications*, was published in four slim issues a year, about a hundred pages each (a pitiful amount for a journal like this, but the publisher refused to give more, so one had to cope), and it was held in extremely high regard around the world. It was translated into English, and many science libraries around the world subscribed to it.

It was very difficult to get a paper published in this journal, partly because of the severe page limitations. There were in fact two types of papers that were published: research articles, each typically ten–fifteen pages long, containing detailed proofs, and short announcements, in which only the results were stated, without proofs. The announcement could not be longer than two pages. In theory, such a short paper was supposed to be followed eventually by a detailed article containing all proofs, but in reality quite often that did not happen because publishing a longer article was extremely difficult. Indeed, it was nearly impossible for a mathematician in the USSR to publish abroad (one needed to get all sorts of security clearances, which could easily take more than a year and a lot of effort). On the other hand, the number of math publications in the Soviet Union, considering the number of mathematicians there, was very small. Unfortunately, many of them were controlled by various groups, which would not allow outsiders to publish, and anti-Semitism was also prevalent in some of them.

Because of all this, a certain subculture of math papers emerged in the USSR, which came to be referred to as the "Russian tradition" of math papers: extremely terse writing, with few details provided. What many mathematicians outside of the Soviet Union did not realize was that this was largely done by necessity, not by choice.

It was this kind of short announcement that Fuchs was aiming at for my first article.

Each article submitted to *Functional Analysis and Applications*, including short announcements, had to be screened and approved by Gelfand. If he liked it, he would then let the article go through the standard refereeing process. This meant that for my article to be considered, I had to meet Israel Moiseevich in person. So before one of the first seminars of the fall semester of 1986, Fuchs introduced me to him.

Gelfand shook my hand, smiled, and said, "I am pleased to meet you. I've heard about you."

I was totally star-struck. I could swear that I saw a halo around Gelfand's head.

Then Gelfand turned to Fuchs and asked him to show my article, which Fuchs handed to him. Gelfand started to turn the pages. There were five of them, which I neatly typed (slowly, with two fingers) on a typewriter I borrowed at Kerosinka, and then inserted formulas by hand.

"Interesting," Gelfand said approvingly, and then turned to Fuchs: "But why is this important?"

Fuchs started to explain something about the discriminant of polynomials of degree n with distinct roots, and how my result could be used to describe the topology of the fiber of the discriminant, and... Gelfand interrupted him:

"Mitya," he said, using the diminutive form of Fuchs' first name, "Do you know how many subscribers the journal has?"

"No, Israel Moiseevich, I don't."

"More than a thousand." That was a pretty large number given how specialized the journal was. "I cannot send you with every issue so that you would explain to each subscriber what this result is good for, now can I?"

Fuchs shook his head.

"This has to be written clearly in the paper, OK?" Gelfand made the point of saying all of this to Fuchs, as though it was all *his* fault. Then he said to both of us: "Otherwise, the paper looks good to me."

With that, he smiled again at me and went to talk to someone else.

Quite an exchange! Fuchs waited until Gelfand was out of the earshot and said to me, "Don't worry about this. He just wanted to impress you." (And he sure did!) "We'll just have to add a paragraph to this effect at the beginning of the paper, and after that he will probably publish it."

That was the best possible outcome. After adding a paragraph required by Gelfand, I officially submitted the article, and eventually it appeared in the journal.[7] With that, my first math project was complete. I crossed my first threshold and was at the beginning of a path that would lead me into the magical world of modern math.

This is the world I want to share with you.

Chapter 7

The Grand Unified Theory

The solution of the first problem was my initiation into the temple of mathematics. Somewhat serendipitously, the next mathematical project I did with Fuchs brought me into the midst of the Langlands Program, one of the deepest and most exciting mathematical theories to emerge in the past fifty years. I will tell you about my project below, but my goal in this book is to describe much more than my own experience. It is to give you a sense of modern math, to prove that it is really about originality, imagination, groundbreaking insights. And the Langlands Program is a great example. I like to think of it as a Grand Unified Theory of Mathematics because it uncovers and brings into focus mysterious patterns shared by different areas of math and thus points to deep, unexpected connections between them.

Mathematics consists of many subfields. They often feel like different continents, with mathematicians working in those subfields speaking different languages. That's why the idea of "unification," bringing together the theories coming from these diverse fields and realizing that they are all part of a single narrative, is so powerful. It's as if you suddenly realized that you could understand another language, one you had desperately tried to learn without much success.

It's useful to think about mathematics as a whole as a giant jigsaw puzzle, in which no one knows what the final image is going to look like. Solving this puzzle is a collective enterprise of thousands of people. They work in groups: here are the algebraists laboring over their part of the puzzle, here are the number theorists, here are the geometers, and so on. Each group has been able to create a small "island" of the big picture, but through most of the history of mathematics, it has been hard to see how these little islands will ever join up.

As a result, most people work on expanding those islands of the puzzle. Every once in a while, however, someone will come who will see how to connect the islands. When this happens, important traits of the big picture emerge, and this gives a new meaning to the individual fields.

This is what Robert Langlands did, but his ambition went deeper than simply joining a few islands. Instead, the Langlands Program that he initiated in the late 1960s has become an attempt to find the mechanism by which we could build bridges between many islands, no matter how unrelated they may seem.

Robert Langlands at his office in Princeton, 1999. Photo by Jeff Mozzochi.

Langlands is now emeritus professor of mathematics at the Institute for Advanced Study in Princeton, where he occupies the office formerly held by Albert Einstein. A man of amazing talent and vision, he was born in 1936 and grew up in a small town near Vancouver; his parents had a millwork business. One of the striking things about Langlands is his fluency in many languages: English, French, German, Russian, and Turkish, even though he didn't speak any languages besides his native English before he entered college.[1]

I have had the opportunity to collaborate with Langlands closely in recent years, and we have often corresponded in Russian. At some point he sent me the list of Russian authors that he had read in the original. The list was so extensive that it seemed he may well have read more of my native Russian literature than I have. I often wonder whether Langlands' unusual language abilities have something to do with his power to bring together different mathematical cultures.

The key point of the Langlands Program is the concept of symmetry that is already familiar to us. We have talked about symmetry in geometry: for example, any rotation is a symmetry of a round table. Our study of these symmetries has led us to the notion of a group. We then saw that groups appear in mathematics in different guises: as groups of rotations, braid groups, and so on. We have also seen that groups were instrumental in classifying elementary particles and predicting the existence of quarks. The groups that are relevant to the Langlands Program appear in the study of numbers.

To explain this, we need to talk first about the numbers that we encounter in our everyday life. Each of us was born in a particular year, lives in a house that has a particular number on the street, has a phone number, a PIN to access a bank account at the ATM, and so forth. All of these numbers have something in common: each of them is obtained by adding number 1 to itself a certain number of times: $1 + 1$ is 2, $1 + 1 + 1$ is 3, and so on. These are called the natural numbers.

We also have the number 0, and the negative numbers: -1, -2, -3,... As we discussed in Chapter 5, these numbers go by the name "integers." So an integer is a natural number, or number 0, or the negative of a natural number.

We also encounter slightly more general numbers. A price, in dollars and cents, is often represented like this: $2.59, meaning two dollars and fifty-nine cents. This is the same as 2 plus the fraction 59/100, or 59 times 1/100. Here 1/100 means the quantity that being added to itself 100 times gives us 1. Numbers of this kind are called rational numbers, or fractions.

A good example of a rational number is a quarter; mathematically, it is represented by the fraction 1/4. More generally, for any two integers m and n we can form the fraction m/n. If m and n have a common divisor, say d (that is to say, $m = dm'$ and $n = dn'$), then we can cancel out d and write m'/n' instead of m/n. For example, 1/4 can also be represented as 25/100, and that's why Americans can say that a quarter is the same thing as 25 cents.

The vast majority of the numbers we encounter in our everyday life situations are these fractions, or rational numbers. But there are also numbers that are not rational. An example is the square root of 2, which we write as follows: $\sqrt{2}$. It is the number whose square is equal to 2. Geometrically, $\sqrt{2}$ is the length of the hypotenuse of the right triangle with legs of length 1.

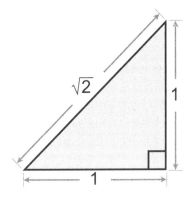

It turns out we cannot represent it as m/n, where m and n are two natural numbers.[2] However, we can approximate it by rational numbers if we write the first few digits of its decimal form: 1.4142, then 1.41421, then 1.414213, and so on. But no matter how many decimal digits we retain, this will be an approximation – there will be more digits to follow. No finite decimal number will ever do justice to $\sqrt{2}$.

Since $\sqrt{2}$ is the length of the hypotenuse of the above triangle, we know that this number is out there. But it just does not fit the numerical system of rational numbers.

There are many other numbers like that, such as $\sqrt{3}$ or the cubic root of 2. We need to develop a systematic way to add these numbers to the rational numbers. Think of the rational numbers as a cup of tea. We can drink it by itself, but our experience will be enhanced if we mix in sugar, milk, honey, various spices – and these are like the numbers $\sqrt{2}$, $\sqrt{3}$, etc.

Let's try to mix in $\sqrt{2}$. This will be the equivalent of adding a cube of sugar to our cup of tea. So we drop $\sqrt{2}$ in the rationals and see what kind of numerical system we obtain. Surely, we want to be able to multiply the numbers within this new numerical system, so we have to include all numbers that are products of rational numbers and $\sqrt{2}$. These have the form $\frac{k}{l}\sqrt{2}$. So our numerical system must include all fractions $\frac{m}{n}$ (these are the rational numbers) and all numbers of the form $\frac{k}{l}\sqrt{2}$. But we also want to be able to add them to each other, so we also have to include the sums

$$\frac{m}{n} + \frac{k}{l}\sqrt{2}.$$

The collection of all numbers of this form is already "self-contained," in the sense that we can perform all the usual operations on them – addition, subtraction, multiplication, and division – and the result will also be a number of the same form.[3] This is our cup of tea with the cube of sugar fully mixed with the tea.

It turns out that this new numerical system has a hidden property that the rational numbers didn't have. This property will be our portal into the magical world of numbers. Namely, it turns out that this numerical system has symmetries.

By a "symmetry" I mean here a rule that assigns a new number to whatever number we begin with. In other words, a given symmetry transforms each number to another number from the same numerical system. We will say that a symmetry is a rule by which each number "goes" to some other number. This rule should be compatible with the operations of addition, subtraction, multiplication, and division. It is not clear yet why we should care about the symmetries of a numerical system. Please bear with me and you will see why momentarily.

Our numerical system has the identity symmetry, the rule by which every number goes to itself. This is like the rotation of a table by 0 degrees, under which every point of the table goes to itself.

It turns out that our numerical system also has a non-trivial symmetry. To explain what it is, let's observe that $\sqrt{2}$ is a solution of the equation $x^2 = 2$. Indeed, if we substitute $\sqrt{2}$ for x, we obtain an equality. But this equation actually has two solutions: one of them is $\sqrt{2}$ and the other is $-\sqrt{2}$. And we have in fact added both of them to the rational numbers when we constructed our new numerical system. Switching these two solutions, we obtain a symmetry of this numerical system.*

To illustrate this more fully in terms of our tea cup analogy, let's modify it slightly. Let's say that we drop a cube of white sugar and a cube of brown sugar in our cup and mix them with the tea. The former is like $\sqrt{2}$ and the latter is like $-\sqrt{2}$. Clearly, exchanging them will not change the resulting cup of tea. Likewise, exchanging $\sqrt{2}$ and $-\sqrt{2}$ will be a symmetry of our numerical system.

Under this exchange, rational numbers remain unchanged.[4] Therefore, the

---

*Note that here and below I use a minus sign (a dash) to represent negative numbers, rather than a hyphen. This conforms to the standard mathematical notation. In fact, there isn't really any difference between the two because $-N = 0 - N$.

number of the form $\dfrac{m}{n} + \dfrac{k}{l}\sqrt{2}$ will go to the number $\dfrac{m}{n} - \dfrac{k}{l}\sqrt{2}$. In other words, in every number we simply change the sign in front of $\sqrt{2}$ and leave everything else the same.[5]

You see, our new numerical system is like a butterfly: the numbers $\dfrac{m}{n} + \dfrac{k}{l}\sqrt{2}$ are like the scales of a butterfly, and the symmetry of these numbers exchanging $\sqrt{2}$ and $-\sqrt{2}$ is like the symmetry of the butterfly exchanging its wings.

More generally, we can consider other equations in the variable x instead of $x^2 = 2$; for example, the cubic equation $x^3 - x + 1 = 0$. If the solutions of such an equation are not rational numbers (as is the case for the above equations), then we can adjoin them to the rational numbers. We can also adjoin to the rational numbers the solutions of several such equations at once. This way we obtain many different numerical systems, or, as mathematicians call them, *number fields*. The word "field" refers to the fact that this numerical system is closed under the operations of addition, subtraction, multiplication, and division.

Just like the number field obtained by adjoining $\sqrt{2}$, general number fields possess symmetries compatible with these operations. The symmetries of a given number field can be applied one after another (composed with each other), just like symmetries of a geometric object. It is not surprising then that these symmetries form a group. This group is called the *Galois group* of the number field,[6] in honor of the French mathematician Évariste Galois.

The story of Galois is one of the most romantic and fascinating stories about mathematicians ever told. A child prodigy, he made groundbreaking discoveries very young. And then he died in a duel at the age of twenty. There are different views on what was the reason for the duel, which happened on May 31, 1832: some say there was a woman involved, and some say it was because of his political activities. Certainly, Galois was uncompromising in expressing his political views, and he managed to upset many people during his short life.

It was literally on the eve of his death that, writing frantically in a candlelit room in the middle of the night, he completed his manuscript outlining his ideas about symmetries of numbers. It was in essence his love letter to humanity in which he shared with us the dazzling discoveries he had made. Indeed, the symmetry groups Galois discovered, which now carry his name, are the wonders of our world, like the Egyptian pyramids or the Hanging Gardens of Babylon. The difference is that we don't have to travel to another continent or through time to find them. They are right at our fingertips, wherever we are. And it's not just their beauty that is captivating; so is their high potency for real-world applications.

Alas, Galois was far ahead of his time. His ideas were so radical that his contemporaries could not understand them at first. His papers were twice rejected by the French Academy of Sciences, and it took almost fifty years for his work to be published and appreciated by other mathematicians. Nevertheless, it is now considered as one of the pillars of modern mathematics.

What Galois had done was bring the idea of symmetry, intuitively familiar to us in geometry, to the forefront of number theory. What's more, he showed symmetry's amazing power.

Before Galois, mathematicians focused on trying to discover explicit formulas for solutions of equations like $x^2 = 2$ and $x^3 - x + 1 = 0$, called polynomial equations. Sadly, this is how we are still taught at school, even though two centuries have passed since Galois' death. For example, we are required to memorize a formula for solutions of a general quadratic equation (that is, of degree 2)

$$ax^2 + bx + c = 0$$

in terms of its coefficients a, b, c. I won't write this formula here so as not to trigger any unpleasant memories. All we need to know about it now is that it involves taking the square root.

Likewise, there is a similar, but more complicated, formula for a general cubic equation (of degree 3)

$$ax^3 + bx^2 + cx + d = 0,$$

in terms of its coefficients a, b, c, d, which involves cubic roots. The task of solving a polynomial equation in terms of radicals (that is, square roots, cubic roots, and so forth) is quickly becoming more and more complicated as the degree of the equation grows.

The general formula for the solutions of the quadratic equations was already known to the Persian mathematician Al-Khwarizmi in the ninth century (the word "algebra" originated from the word "al-jabr," which appears in the title of his book). Formulas for solutions of the cubic and quartic (degree 4) equations were discovered in the first half of the sixteenth century. Naturally, the next target was a quintic equation (of degree 5). Prior to Galois, many mathematicians had been desperately trying to find a formula for its solutions for almost 300 years, to no avail. But Galois realized that they had been asking the wrong question. Instead, he said, we should focus on the group of symmetries of the number field obtained by adjoining the solutions of this equation to the rational numbers – this is what we now call the Galois group.

The question of describing the Galois group turns out to be much more tractable than the question of writing an explicit formula for the solutions. One can say something meaningful about this group even without knowing what the solutions are. And from this one can then infer important information about the solutions. In fact, Galois was able to show that a formula for solutions in terms of radicals (that is, square roots, cubic roots, and so on) exists if and only if the corresponding Galois group has a particularly simple structure: is what mathematicians now call a *solvable* group. For quadratic, cubic, and quartic equations, the Galois groups are always solvable. That's why solutions of these equations may be written in terms of the radicals. But Galois showed that the group of symmetries of a typical quintic equation (or an equation of a higher degree) is not solvable. This immediately implies that there is no formula for solutions of these equations in terms of radicals.[7]

I won't get into the details of this proof, but let's consider a couple of examples of Galois groups to give you an idea what these groups look like. We have already described the Galois group in the case of the equation $x^2 = 2$. This equation has two solutions, $\sqrt{2}$ and $-\sqrt{2}$, which we adjoin to the rational numbers. The Galois group of the resulting number field[8] then consists of two elements: the identity and the symmetry exchanging $\sqrt{2}$ and $-\sqrt{2}$.

As our next example, consider a cubic equation written above, and suppose that its coefficients are rational numbers, but all of its three solutions are irrational. We then construct a new number field by adjoining these solutions to the rational numbers. It's like adding three different ingredients to our cup of tea: say, a cube of sugar, a dash of milk, and a spoonful of honey. Under any symmetry of this number field (the cup of tea with these ingredients added), the cubic equation won't change because its coefficients are rational numbers, which are preserved by symmetries. Hence each solution of the cubic equation (one of the three ingredients) will necessarily go to another solution. This observation allows us to describe the Galois group of symmetries of this number field in terms of permutations of these three solutions. The main point is that we obtain this description without writing down any formulas for the solutions.[9]

Similarly, the Galois group of symmetries of the number field obtained by adjoining all solutions of an arbitrary polynomial equation to the rational numbers may also be described in terms of permutations of these solutions (there will be n solutions for a polynomial equation of degree n whose solutions are all distinct and not rational). This way we can infer a lot of information about the equation without expressing its solutions in terms of the coefficients.[10]

Galois' work is a great example of the power of a mathematical insight. Galois did not solve the problem of finding a formula for solutions of polynomial equations in the sense in which it was understood. He *hacked* the problem! He reformulated it, bent and warped it, looked at it in a totally different light. And his brilliant insight has forever changed the way people think about numbers and equations.

And then, 150 years later, Langlands took these ideas much farther. In 1967, he came up with revolutionary insights tying together the theory of Galois groups and another area of mathematics called harmonic analysis. These two areas, which seem light years apart, turned out to be closely related. Langlands, then in his early thirties, summarized his ideas in a letter to the eminent mathematician André Weil. Copies were widely circulated among mathematicians at the time.[11] The letter's cover note is remarkable for its understatement:[12]

> Professor Weil: In response to your invitation to come and talk, I wrote the enclosed letter. After I wrote it I realized there was hardly a statement in it of which I was certain. If you are willing to read it as pure speculation I would appreciate that; if not – I am sure you have a waste basket handy.

What followed was the beginning of a groundbreaking theory that forever changed the way we think about mathematics. Thus, the Langlands Program was born.

Several generations of mathematicians have dedicated their lives to solving the problems put forward by Langlands. What was it that so inspired them? The answer is coming up in the next chapter.

Chapter 8

Magic Numbers

When we first talked about symmetries in Chapter 2, we saw that representations of a group named $SU(3)$ govern the behavior of elementary particles. The focus of the Langlands Program is also on representations of a group, but this time it is the Galois group of symmetries of a number field of the kind discussed in the previous chapter. It turns out that these representations form the "source code" of a number field, carrying all essential information about numbers.

Langlands' marvelous idea was that we can extract this information from objects of an entirely different nature: the so-called automorphic functions, which come from another field of mathematics called harmonic analysis. The roots of harmonic analysis are in the study of harmonics, which are the basic sound waves whose frequencies are multiples of each other. The idea is that a general sound wave is a superposition of harmonics, the way the sound of a symphony is a superposition of the harmonics corresponding to the notes played by various instruments. Mathematically, this means expressing a given function as a superposition of the functions describing harmonics, such as the familiar trigonometric functions sine and cosine. Automorphic functions are more sophisticated versions of these familiar harmonics. There are powerful analytic methods for doing calculations with these automorphic functions. And Langlands' surprising insight was that we can use these functions to learn about much more difficult questions in number theory. This way we find a hidden harmony of numbers.

I wrote in the Preface that one of the principal functions of mathematics is the ordering of information, or, as Langlands himself put it, "creating order from seeming chaos."[1] Langlands' idea is so powerful precisely because it helps organize seemingly chaotic data from number theory into regular patterns full of symmetry and harmony.

If we think of different fields of mathematics as continents, then number theory would be like North America and harmonic analysis like Europe. Over the years, it's been taking us less and less time to travel from one continent to the other. It used to take days by boat, and now only hours by plane. But imagine that a new technology was invented that would allow you to be instantly transported from anywhere in North America to someplace in Europe. That would be an equivalent of the connections discovered by Langlands.

I will now describe one of these breathtaking connections, which is closely related to Fermat's Last Theorem that we talked about in Chapter 6.

Fermat's Last Theorem is deceptively simple to state. It says that there are no natural numbers x, y, and z solving the equation

$$x^n + y^n = z^n,$$

if n is greater than 2.

As I wrote, this result was guessed by the French mathematician Pierre Fermat more than 350 years ago, in 1637. He wrote about it on the margin of an old book he was reading, saying that he had found a "truly marvelous" proof of this statement, but "the margin is too small to contain it." Call it a seventeenth-century Twitter-style proof: "I have found a marvelous proof of this theorem, but unfortunately I can't write it here because it's longer than one hundred and forty chara" – sorry, ran out of space.

There is little doubt that Fermat was mistaken. It took more than 350 years to find the real proof, and it is incredibly complicated. There are two main steps: first, in 1986, Ken Ribet showed that Fermat's Last Theorem follows from the so-called Shimura–Taniyama–Weil conjecture.

(Perhaps, I should note that a mathematical conjecture is a statement that one expects to be true, but for which one does not yet know a proof. Once the proof is found, the conjecture becomes a theorem.[2])

What Ken Ribet showed was that if there exist natural numbers x, y, z solving Fermat's equation, then, using these numbers, one can construct a certain cubic equation, which has a property precluded by the Shimura–Taniyama–Weil conjecture (I will explain below what this equation and this property are). If we know that the Shimura–Taniyama–Weil conjecture is true, then this equation cannot exist. But then the numbers x, y, z solving Fermat's equation cannot exist either.[3]

Let's pause for a minute and go over the logic of this argument one more time. In order to prove Fermat's Last Theorem, we assume that it is false; that is, we suppose that there exist natural numbers x, y, z such that Fermat's equation is satisfied. Then we associate to these numbers a cubic equation, which turns out to have a certain undesirable property. The Shimura–Taniyama–Weil conjecture tells us that such an equation *cannot exist*. But then these numbers x, y, z cannot exist either. Hence there can be no solutions to Fermat's equation. Therefore Fermat's Last Theorem is true! Schematically, the flow chart of this argument looks as follows (we abbreviate "Fermat's Last Theorem" as FLT and "Shimura–Taniyama–Weil conjecture" as STWC):

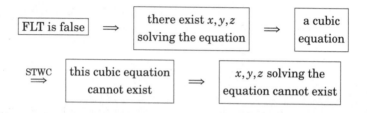

This kind of argument is called *proof by contradiction*. We start with the statement that is opposite to what we are trying to prove (in our case, it is the statement that there exist natural numbers x, y, z solving Fermat's equation, which is opposite to what we want to prove). If, through a chain of implications, we then arrive at a statement that is demonstrably false (in our case, the existence of a cubic equation that is prohibited by the Shimura–Taniyama–Weil conjecture), then we conclude that the statement we started with is false. Hence the statement we wanted to prove (Fermat's Last Theorem) is true.

What remains then to establish Fermat's Last Theorem is to prove that the Shimura–Taniyama–Weil conjecture is true. Once this was understood (in 1986, after Ribet's work), the search was on for a proof of the Shimura–Taniyama–Weil conjecture.

Several proofs had been announced over the years, but subsequent analysis showed that these proofs contained mistakes or gaps. In 1993, Andrew Wiles claimed that he had proved the conjecture, but a few months later it was found that there was a gap in his proof. For a while, it looked like his proof would be remembered alongside many other famous "non-proofs," in which gaps were found, but never closed.

Luckily, Wiles was able to close the gap within a year, with the help of another mathematician, Richard Taylor. Together, they completed the proof.[4] In a wonderful documentary film about Fermat's Last Theorem, Wiles gets emotional when he recounts this moment, and one can only imagine what a gut-wrenching experience it must have been for him.

Thus, the Shimura–Taniyama–Weil conjecture is a key result in proving Fermat's Last Theorem. It may also be viewed as a special case of the Langlands Program, and hence it provides an excellent illustration of the unexpected connections predicted by the Langlands Program.

The Shimura–Taniyama–Weil conjecture is a statement about certain equations. A large part of mathematics is in fact about solving equations. We want to know whether a given equation has a solution in a given domain; if so, can we find one? If there are several solutions, then how many? Why do some equations have solutions and some don't?

In the previous chapter we talked about polynomial equations on one variable, such as $x^2 = 2$. Fermat's Last Theorem is about an equation on three variables: $x^n + y^n = z^n$. And the Shimura–Taniyama–Weil conjecture is about a class of algebraic equations on two variables, such as this one:

$$y^2 + y = x^3 - x^2.$$

A solution of this equation is a pair of numbers x, y such that the left-hand side is equal to the right-hand side.

But what kind of numbers do we want x and y to be? There are several choices: one possibility is to say that x and y are natural numbers or integers. Another possibility is to say rational numbers. We can also look for solutions x, y that are real numbers, or even complex numbers – we will discuss this option in more detail in the next chapter.

It turns out that there is one more choice, which is less obvious but equally important: to consider solutions x, y "modulo N," for some fixed natural number N. This means that we look for integers x and y such that the left-hand side is equal to the right-hand side up to a number that is divisible by N.

For example, let's look for solutions modulo $N = 5$. There is one obvious solution $x = 0, y = 0$. And there are three other, slightly less obvious, solutions: $x = 0, y = 4$ is a solution modulo 5 because the left-hand side is then 20 and the right-hand side is 0. The difference between the left- and right-hand sides is 20, which is divisible by 5. So this is indeed a solution of the equation modulo 5. By a similar argument, $x = 1, y = 0$ and $x = 1, y = 4$ are also solutions modulo 5.

We have already discussed this kind of arithmetic in Chapter 2 when we talked about the group of rotations of a round table. We saw then that the addition of angles was done "modulo 360." That is to say, if the result of addition of two angles is greater than 360 degrees, we subtract from it 360 to bring it into the range from 0 to 360. For example, rotation by 450 degrees is the same as rotation by 90 degrees, because $450 - 360 = 90$.

We also encounter this arithmetic when we use the clock. If we start working at 10 o'clock in the morning and work for 8 hours, when do we finish? Well, $10 + 8 = 18$, so a natural thing to say would be: "We finish at 18 o'clock." This would be perfectly fine to say in France where they record hours as numbers from 0 to 24 (actually, not so fine, because a working day in France is usually limited to seven hours). But in the U.S. we say: "We finish at 6 pm." How do we get 6 out of 18? We subtract 12 from it: $18 - 12 = 6$.

So we use the same idea with hours as we do with angles. In the first case, we do addition "modulo 360." In the second case, we do addition "modulo 12."

Likewise, we can do addition modulo any natural number N. Consider the set of all consecutive whole numbers between 0 and $N - 1$,

$$\{0, 1, 2, \dots, N - 2, N - 1\}.$$

If $N = 12$, this is the set of possible hours. In general, the role of 12 is played by number N, so that it's not 12 that takes us back to 0, but N.

We define addition on the set of these numbers in the same way as for the hours. Given any two numbers from this set, we add them up, and if the result is greater than N, we subtract N from it to get a number from the same set. This operation makes this set into a group. The identity element is the number 0: adding it to any other number does not change it. Indeed, we have $n + 0 = n$. And for any number n from our set, its "additive inverse" is $N - n$, because $n + (N - n) = N$, which is the same as 0 according to our rules.

For example, let's take $N = 3$. Then we have the set $\{0, 1, 2\}$ and addition modulo 3. For example, we have

$$2 + 2 = 1 \qquad \text{modulo} \quad 3$$

in this system, because $2 + 2 = 4$, but since $4 = 3 + 1$, the number 4 is equal to 1 modulo 3.

So if someone says to you: "2 plus 2 equals 4" to indicate a well-established fact, you can now say (with a condescending smile if you like): "Well, actually, that's not always true." And if they ask you to explain what you mean, you can tell them, "If you do addition modulo 3, then 2 plus 2 is equal to 1."

Given any two numbers from the above set, we can also multiply them. The result may not be between 0 and $N - 1$, but there will be a unique number in this range that will differ from the result of multiplication by something divisible by N. However, in general, the set $\{1, 2, ..., N - 1\}$ is not a group with respect to multiplication. We do have the identity element: number 1. But not every element has the multiplicative inverse modulo N. This happens if and only if N is a *prime number*, that is, a number that is not divisible by any other natural number other than 1 and itself.[5]

The first few primes are 2, 3, 5, 7, 11, 13,... (It is customary to exclude 1 from this list.) Even natural numbers, except for 2, are not prime, because they are divisible by 2, and 9 isn't prime, because it is divisible by 3. There are in fact infinitely many primes – no matter how large a prime number is, there is another prime number that is even larger.[6] Primes, because they are indivisible, are the elementary particles of the world of natural numbers; every other natural number, in fact, can be written, in a unique way, as the product of prime numbers. For example, $60 = 2 \cdot 2 \cdot 3 \cdot 5$.

Let us fix a prime number. As is customary, we will denote it by p. Then we consider the set of all consecutive whole numbers between 0 to $p - 1$; that is,

$$\{0, 1, 2, 3, 4, ..., p - 2, p - 1\}.$$

And we consider two operations on them: addition and multiplication modulo p.

As have seen above, this set is a group with respect to addition modulo p. What is even more remarkable is that if we remove number 0 and consider the set of consecutive whole numbers between 1 and $p - 1$, that is $\{1, 2, ..., p - 1\}$, we obtain a group with respect to multiplication modulo p. The element 1 is the multiplicative identity (this is clear), and I claim that any natural number between 1 and $p - 1$ has a multiplicative inverse.[7]

For example, if $p = 5$, we find that

$$2 \cdot 3 = 1 \qquad \text{modulo} \quad 5,$$

and

$$4 \cdot 4 = 1 \qquad \text{modulo} \quad 5,$$

so that the multiplicative inverse of 2 modulo 5 is 3, and 4 is its own inverse modulo 5. It turns out that this is true in general.[8]

In our everyday life, we are used to numbers that are integers or fractions. Sometimes, we use numbers like $\sqrt{2}$. But we have now discovered a numerical system of an entirely different nature: the finite set of numbers $\{0, 1, 2, \ldots, p-1\}$, where p is a prime, on which we have the operations of addition and multiplication modulo p. It is called the *finite field* with p elements. These finite fields form an important archipelago in the world of numbers – one that, unfortunately, most of us are never told exists.

Even though these numerical systems look very different from the numerical systems we are used to, such as rational numbers, they have the same salient properties: they are closed under the operations of addition, subtraction, multiplication, and division.[9] Therefore, everything we can do with the rational numbers may also be done with these, more esoteric looking, finite fields.

Actually, they are not so esoteric any more, having found important applications – most notably, in cryptography. When we make a purchase online and enter our credit card number, this number gets encrypted using the arithmetic modulo primes, which is dictated by the equations very much like the ones we have looked at above (see the description of the RSA encryption algorithm in endnote 7 to Chapter 14).

Let's go back to the cubic equation

$$y^2 + y = x^3 - x^2$$

that we considered above. Let us look for solutions of this equation modulo p, for various primes p. For example, we have seen above that there are 4 solutions modulo 5. But note that the solutions modulo $p = 5$ are not necessarily solutions modulo other primes (say, $p = 7$ or $p = 11$). So these solutions do depend on the prime p modulo which we do the arithmetic.

The question we are going to ask now is the following: how does the number of solutions of this equation, taken modulo p, depend on p? For small p, we can count them explicitly (perhaps, with the aid of a computer), so we can actually compile a small table.

Mathematicians have known for some time that the number of solutions of an equation of this type modulo p is roughly equal to p. Let's denote the "deficit," the number by which the actual number of solutions differs from the expected number of solutions (namely, p), by a_p. This means that the number of solutions of the above equation modulo p is equal to $p - a_p$. The numbers a_p

could be positive or negative for a given p. For example, we found above that for $p = 5$ there are 4 solutions. Since $4 = 5 - 1$, we obtain that $a_5 = 1$.

We can find the numbers a_p for small primes on a computer. They seem to be random. There does not appear to be any natural formula or rule that would enable us to compute them. What's worse, the computation very quickly becomes immensely complicated.

But what if I told you that there was in fact a simple rule that generated the numbers a_p all at once?

In case you are wondering what exactly I mean here by a "rule" generating these numbers, let's consider a more familiar sequence, the so-called Fibonacci numbers:

$$1, 1, 2, 3, 5, 8, 13, 21, 34, \ldots$$

Named after an Italian mathematician who introduced them in his book published in 1202 (in the context of a problem of mating rabbits, no less), Fibonacci numbers are ubiquitous in nature: from petal arrangements in flowers to the patterns on the surface of a pineapple. They also have many applications, such as the "Fibonacci retracement" in the technical analysis of stock trading.

The Fibonacci numbers are defined as follows: the first two of them are equal to 1. Each number after that is equal to the sum of the preceding two Fibonacci numbers. For example, $2 = 1 + 1, 3 = 2 + 1, 5 = 3 + 2$, and so on. If we denote the nth Fibonacci number by F_n, then we have $F_1 = 1, F_2 = 1$ and

$$F_n = F_{n-1} + F_{n-2}, \qquad n > 2.$$

In principle, this rule enables us to find the nth Fibonacci number for any n. But in order to do this, we have to first find all Fibonacci numbers F_i for i between 1 and $n - 1$.

However, it turns out that these numbers could also be generated in the following way. Consider the series

$$q + q(q + q^2) + q(q + q^2)^2 + q(q + q^2)^3 + q(q + q^2)^4 + \ldots.$$

In words, we multiply an auxiliary variable q by the sum of all powers of the expression $(q + q^2)$. If we open the brackets, we obtain an infinite series, whose first terms are

$$q + q^2 + 2q^3 + 3q^4 + 5q^5 + 8q^6 + 13q^7 + \ldots$$

For example, let's compute the term with q^3. It can only occur in q, $q(q + q^2)$, and $q(q + q^2)^2$. (Indeed, all other expressions that appear in the defining sum, such as $q(q + q^2)^3$, will only contain powers of q greater than 3.) The first of these does not contain q^3, and each of the other two contains q^3 once. Their sum yields $2q^3$. We obtain in a similar way other terms of the series.

Analyzing the first terms of this series, we find that for n between 1 and 7, the coefficient in front of q^n is the nth Fibonacci number F_n. For example, we have the term $13q^7$ and $F_7 = 13$. It turns out that this is true for all n. For this reason, mathematicians call this infinite series the *generating function* of the Fibonacci numbers.

This remarkable function can be used to give an effective formula for calculating the nth Fibonacci number without any reference to the preceding Fibonacci numbers.[10] But even putting the computational aspects aside, we can appreciate the value added by this generating function: instead of giving a self-referential recursive procedure, the generating function beholds all Fibonacci numbers at once.

Let's go back to the numbers a_p counting the solutions of the cubic equation modulo primes. Think of these numbers as analogues of the Fibonacci numbers (let's ignore the fact that the numbers a_p are labeled by the prime numbers p, whereas the Fibonacci numbers F_n are labeled by all natural numbers n).

It seems nearly unbelievable that there would be a rule generating these numbers. And yet, German mathematician Martin Eichler discovered one in 1954.[11] Namely, consider the following generating function:

$$q(1 - q)^2(1 - q^{11})^2(1 - q^2)^2(1 - q^{22})^2(1 - q^3)^2(1 - q^{33})^2(1 - q^4)^2(1 - q^{44})^2 \ldots$$

In words, this is q times the product of factors of the form $(1 - q^a)^2$, with a going over the list of numbers of the form n and $11n$, where $n = 1, 2, 3, \ldots$. Let's open the brackets, using the standard rules:

$$(1 - q)^2 = 1 - 2q + q^2, \qquad (1 - q^{11})^2 = 1 - 2q^{11} + q^{22}, \qquad \ldots$$

and then multiply all the factors. Collecting the terms, we obtain an infinite sum, which begins like this:

$$q - 2q^2 - q^3 + 2q^4 + q^5 + 2q^6 - 2q^7 - 2q^9 - 2q^{10} + q^{11} - 2q^{12} + 4q^{13} + \ldots$$

and the ellipses stand for the terms with the powers of q greater than 13. Though this series is infinite, each coefficient is well-defined because it is determined by finitely many factors in the above product. Let us denote the coefficient in front of q^m by b_m. So we have $b_1 = 1, b_2 = -2, b_3 = -1, b_4 = 2, b_5 = 1$, etc. It is easy to compute them by hand or on a computer.

An astounding insight of Eichler was that for all prime numbers p, the coefficient b_p is equal to a_p. In other words, $a_2 = b_2, a_3 = b_3, a_5 = b_5, a_7 = b_7$, and so on.

Let's check, for example, that this is true for $p = 5$. In this case, looking at the generating function we find that the coefficient in front of q^5 is $b_5 = 1$. On the other hand, we have seen that our cubic equation has 4 solutions modulo $p = 5$. Therefore $a_5 = 5 - 4 = 1$, so indeed $a_5 = b_5$.

We started out with what looked like a problem of infinite complexity: counting solutions of the cubic equation

$$y^2 + y = x^3 - x^2$$

modulo p, for all primes p. And yet, all information about this problem is contained in a single line:

$$q(1-q)^2(1-q^{11})^2(1-q^2)^2(1-q^{22})^2(1-q^3)^2(1-q^{33})^2(1-q^4)^2(1-q^{44})^2 \ldots$$

This one line is a secret code containing all information about the numbers of solutions of the cubic equation modulo all primes.

A useful analogy would be to think of the cubic equation like a sophisticated biological organism, and its solutions as various traits of this organism. We know that all of these traits are encoded in the DNA molecule. Likewise, all the complexity of our cubic equation turns out to be encoded in a generating function, which is like the DNA of this equation. Furthermore, this function is defined by a simple rule.

What's even more fascinating is that if q is a number whose absolute value is less than 1, then the above infinite sum converges to a well-defined number. So we obtain a function in q, and this function turns out to have a very special property that is similar to the periodicity of the familiar trigonometric functions, sine and cosine.

The sine function $\sin(x)$ is periodic with the period 2π, that is to say, $\sin(x + 2\pi) = \sin(x)$. But then also $\sin(x+4\pi) = \sin(x)$, and more generally $\sin(x+2\pi n) = \sin(x)$ for any integer n. Think about it this way: each integer n gives rise to a symmetry of the line: every point x on the line is shifted to $x + 2\pi n$. Therefore, the group of all integers is realized as a group of symmetries of the line. The periodicity of the sine function means that this function is invariant under this group.

Likewise, the Eichler generating function of the variable q written above turns out to be invariant under a certain symmetry group. Here we should take q to be not a real, but rather a complex number (we will discuss this topic in the next chapter). Then we can view q not as a point on the line, as in the case of the sine function, but as a point inside a unit disc on the complex plane. The symmetry property is similar: on this disc there is a group of symmetries, and our function is invariant under this group.[12] A function with this kind of invariance property is called a *modular form*.

This symmetry group of the disc is very rich. To get an idea of what it is, let's look at this picture, on which the disc is broken into infinitely many triangles.[13]

The symmetries act on the disc by exchanging these triangles. In fact, for any two triangles, there is a symmetry exchanging them. Though these symmetries of the disc are quite sophisticated, this is analogous to how, when the group of integers acts on the line, its symmetries move around the intervals $[2\pi m, 2\pi(m+1)]$. The sine function is invariant under those symmetries, whereas the Eichler generating function is invariant under symmetries of the disc.

As I mentioned at the beginning of this chapter, the sine function is the simplest example of a "harmonic" (basic wave) that is used in the harmonic analysis on the line. Likewise, the Eichler function, together with other modular forms, are the harmonics that appear in the harmonic analysis on the unit disc.

The magnificent insight of Eichler was that the seemingly random numbers of solutions of a cubic equation modulo primes come from a single generating function, which obeys an exquisite symmetry – revealing a hidden harmony and order in those numbers. Similarly, as if in a stroke of black magic, the Langlands Program organizes previously inaccessible information into regular patterns, weaving a delicate tapestry of numbers, symmetries, and equations.

When I first talked about mathematics at the beginning of this book, you may have wondered what I meant by a mathematical result being "beautiful" or "elegant." This is what I meant. The fact that these highly abstract notions coalesce in such refined harmony is absolutely mind-boggling. It points to something rich and mysterious lurking beneath the surface, as if the curtain had been lifted and we caught glimpses of the reality that had been carefully hidden from us. These are the wonders of modern math, and of the modern world.

One might also ask whether, in addition to possessing innate beauty and establishing a surprising link between areas of mathematics that seem to be far removed from each other, this result has any practical applications. This is a fair question. At present, I am not aware of any. But cubic equations over finite fields of p elements of the kind we have considered above (which give rise to the so-called elliptic curves) are widely used in cryptography.[14] So I would not be surprised if the analogues of Eichler's result will also one day find applications as powerful and ubiquitous as encryption algorithms.

The Shimura–Taniyama–Weil conjecture is a generalization of Eichler's result. It says that for *any* cubic equation like the one above (subject to some mild conditions), the numbers of solutions modulo primes are the coefficients of a modular form. Moreover, there is a one-to-one correspondence between the cubic equations and the modular forms of a certain kind.

What do I mean here by a one-to-one correspondence? Suppose that we have five pens and five pencils. We can assign a pencil to each pen in such a way that each pencil is assigned to one and only pen. This is called a one-to-one correspondence.

There are many different ways to do it. But suppose that under our one-to-one correspondence each pen has exactly the same length as the pencil assigned to it. We will then call the length an "invariant" and say that our correspondence preserves this invariant. If all pens have different lengths, the one-to-one correspondence will be uniquely determined by this property.

Now, in the case of the Shimura–Taniyama–Weil conjecture, the objects on one side are the cubic equations such as the one above. These will be our pens, and for each of them the numbers a_p will be the invariants attached to it. (It's like the length of a pen, except that now there isn't just one invariant, but a whole collection labeled by primes p.)

The objects on the other side of the correspondence are modular forms. These will be our pencils, and for each of them, the coefficients b_p will be the invariants attached to it (like the length of a pencil).

The Shimura–Taniyama–Weil conjecture says that there is a one-to-one correspondence between these objects preserving these invariants:

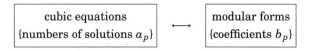

That is to say, for any cubic equation there exists a modular form such that $a_p = b_p$ for all primes p, and vice versa.[15]

Now I can explain the link between the Shimura–Taniyama–Weil conjecture and Fermat's Last Theorem: starting from a solution of the Fermat equation, we can construct a certain cubic equation.[16] However, Ken Ribet showed that the numbers of solutions of this cubic equation modulo primes cannot be the coefficients of a modular form whose existence is stipulated by the Shimura–Taniyama–Weil conjecture. Once this conjecture is proved, we conclude that such a cubic equation cannot exist. Therefore, there are no solutions to the Fermat equation.

The Shimura–Taniyama–Weil conjecture is a stunning result because the numbers a_p come from the study of solutions of an equation modulo primes – they are from the world of number theory – and the numbers b_p are the coefficients of a modular form, from the world of harmonic analysis. These two worlds seem to be light years apart, and yet it turns out that they describe one and the same thing!

The Shimura–Taniyama–Weil conjecture may be recast as a special case of the Langlands Program. In order to do that, we replace each of the cubic equations appearing in the Shimura–Taniyama–Weil conjecture by a certain two-dimensional representation of the Galois group. This representation is naturally obtained from the cubic equation, and the numbers a_p can be attached directly to this representation (rather than the cubic equation). Therefore the conjecture may be expressed as a relation between two-dimensional representations of the Galois group and modular forms.

(I recall from Chapter 2 that a two-dimensional representation of a group is a rule that assigns a symmetry of a two-dimensional space (that is, a plane) to each element of this group. For example, in Chapter 2 we talked about a two-dimensional representation of the circle group.)

Even more generally, conjectures of the Langlands Program relate, in unexpected and profound ways, n-dimensional representations of the Galois group (which generalize the two-dimensional representations corresponding to the cubic equations in the Shimura–Taniyama–Weil conjecture) and the so-called *automorphic functions* (which generalize the modular forms in the Shimura–Taniyama–Weil conjecture):

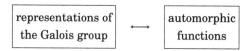

Though there is little doubt that these conjectures are true, most of them are still unproved to this day, despite an enormous effort by several generations of mathematicians in the past forty-five years.

You may be wondering: how could one come up with these kinds of conjectures in the first place?

This is really a question about the nature of mathematical insight. The ability to see patterns and connections that no one had seen before does not come easily. It is usually the product of months, if not years, of hard work. Little by little, the inkling of a new phenomenon or a theory emerges, and at first you don't believe it yourself. But then you say: "what if it's true?" You try to test the idea by doing sample calculations. Sometimes these calculations are hard, and you have to navigate through mountains of heavy formulas. The probability of making a mistake is very high, and if it does not work at first, you try to redo it, over and over again.

More often than not, at the end of the day (or a month, or a year), you realize that your initial idea was wrong, and you have to try something else. These are the moments of frustration and despair. You feel that you have wasted an enormous amount of time, with nothing to show for it. This is hard to stomach. But you can never give up. You go back to the drawing board, you analyze more data, you learn from your previous mistakes, you try to come up with a better idea. And every once in a while, suddenly, your idea starts to work. It's as if you had spent a fruitless day surfing, when you finally catch a wave: you try to hold on to it and ride it for as long as possible. At moments like this, you have to free your imagination and let the wave take you as far as it can. Even if the idea sounds totally crazy at first.

The statement of the Shimura–Taniyama–Weil conjecture must have sounded crazy to its creators. How could it not? Yes, the conjecture had its roots in earlier results, such as those by Eichler that we discussed above (which were subsequently generalized by Shimura), which showed that for *some* cubic equations, the numbers of solutions modulo p were recorded in the coefficients of a modular form. But the idea that this was true for *any* cubic equation must have sounded totally outrageous at the time. This was a leap of faith, first made by the Japanese mathematician Yutaka Taniyama, in the form of a question that he posed at the International Symposium on Algebraic Number Theory held in Tokyo in September 1955.

I've always wondered: what did it take for him to come to *believe* that this wasn't crazy, but real? To have the courage to say it publicly?

We'll never know. Unfortunately, not long after his great discovery, in November 1958, Taniyama committed suicide. He was only thirty-one. To add to the tragedy, shortly afterward the woman whom he was planning to marry also took her life, leaving the following note:[17]

> We promised each other that no matter where we went, we would never be separated. Now that he is gone, I must go too in order to join him.

The conjecture was subsequently made more precise by Taniyama's friend and colleague Goro Shimura, another Japanese mathematician. Shimura has worked most of his life at Princeton University and is currently an emeritus professor there. He has made major contributions to mathematics, many pertinent to the Langlands Program, and several fundamental concepts in this area carry his name (such as the "Eichler–Shimura congruence relations" and "Shimura varieties").

In his thoughtful essay about Taniyama, Shimura made this striking comment:[18]

> Though he was by no means a sloppy type, he was gifted with the special capability of making many mistakes, mostly in the right direction. I envied him for this, and tried in vain to imitate him, but found it quite difficult to make good mistakes.

In the words of Shimura, Taniyama "wasn't very careful when he stated his problem" at the Symposium in Tokyo in September 1955.[19] Some corrections had to be made. And yet, this was a revolutionary insight, which led to one of the most significant achievements in mathematics of the twentieth century.

The third person whose name is attached to the conjecture is André Weil, whom I have mentioned earlier. He is one of the giants of mathematics in the twentieth century. Known for his brilliance as well as his temper, he was born in France and came to the United States during World War II. After holding academic appointments at various American universities, he settled at the Institute for Advanced Study in Princeton in 1958 and stayed there until his death in 1998, at age 92.

André Weil, 1981. Photo by Herman Landshoff. From the Shelby White and Leon Levy Archives Center, Institute for Advanced Study, Princeton.

Weil is particularly relevant to the Langlands Program, and not just because the famous letter in which Robert Langlands first formulated his ideas was addressed to him, or because of the Shimura–Taniyama–Weil conjecture. The Langlands Program is best seen through the prism of the "big picture" of mathematics that André Weil outlined in a letter to his sister. We will talk about it in the next chapter. This will be our stepping stone toward bringing the Langlands Program into the realm of geometry.

Chapter 9

Rosetta Stone

In 1940, during the war, André Weil was imprisoned in France for refusing to serve in the army. As the obituary published in *The Economist* put it,[1]

> [Weil] had been deeply struck.... by the damage wreaked upon mathematics in France by the first world war, when "a misguided notion of equality in the face of sacrifice" led to the slaughter of the country's young scientific elite. In the light of this, he believed he had a duty, not just to himself but also to civilization, to devote his life to mathematics. Indeed, he argued, to let himself be diverted from the subject would be a sin. When others raised the objection "but if everybody were to behave like you...", he replied that this possibility seemed to him so implausible that he did not feel obliged to take it into account.

While in prison, Weil wrote a letter to his sister Simone Weil, a famous philosopher and humanist. This letter is a remarkable document; in it, he tries to explain in fairly elementary terms (accessible even to a philosopher – just kidding!) the "big picture" of mathematics as he saw it. Doing so, he set a great example to follow for all mathematicians. I sometimes joke that perhaps we should jail some of the leading mathematicians to force them to express their ideas in accessible terms, the way Weil did.

Weil writes in the letter about the role of analogy in mathematics, and he illustrates it by the analogy that interested him the most: between number theory and geometry.

This analogy proved to be extremely important for the development of the Langlands Program. As we discussed earlier, the roots of the Langlands Program are in number theory. Langlands conjectured that hard questions of number theory, such as the counting of solutions of equations modulo primes, can be solved by using methods of harmonic analysis – more specifically, the study of automorphic functions. This is exciting: first of all, it gives us a new way to solve what previously looked like intractable problems. And second, it points to deep and fundamental connections between different areas of mathematics. So naturally, we want to know what is really going on here: why might these hidden connections exist? And we still don't fully understand it. Even the Shimura–Taniyama–Weil conjecture took a very long time to be resolved. And it's only a special case of the general Langlands conjectures. There are hundreds and thousands of similar statements that are still not proved.

So how should we approach these difficult conjectures? One way is just to keep working hard and try to come up with new ideas and insights. This has been happening, and significant progress has been made. Another possibility is to try to expand the scope of the Langlands Program. Since it points to some essential structures in number theory and harmonic analysis and connections between them, chances are that similar structures and connections can also be found between other fields of mathematics.

This has indeed turned out to be the case. It was gradually realized that the same mysterious patterns may be observed in other areas of mathematics, such as geometry, and even in quantum physics. When we learn something about these patterns in one area, we get hints about their meaning in other areas. I have written earlier that the Langlands Program is a Grand Unified Theory of mathematics. What I mean by this is that the Langlands Program points to some universal phenomena and connections between these phenomena across different fields of mathematics. And I believe that it holds the keys to understanding what mathematics is really about, far beyond the original Langlands conjectures.

The Langlands Program is now a vast subject. There is a large community of people working on it in different fields: number theory, harmonic analysis, geometry, representation theory, mathematical physics. Although they work with very different objects, they are all observing similar phenomena. And these phenomena give us clues to understanding how these diverse domains are interconnected, like parts of a giant jigsaw puzzle.

My entry point to the Langlands Program was through my work on Kac–Moody algebras, which I will describe in detail in the next few chapters. But the more I learned about the Langlands Program, the more I got excited by how ubiquitous it is in mathematics.

Think of different areas of modern math as languages. We have sentences from these languages that we think mean the same thing. We put them next to each other, and little by little we start developing a dictionary that allows us to translate between different areas of mathematics. André Weil gave us a suitable framework for understanding connections between number theory and geometry, a kind of "Rosetta stone" of modern math.

On one side, we have objects of number theory: rational numbers and other number fields that we discussed in the previous chapter, such as the one obtained by adjoining $\sqrt{2}$, and their Galois groups.

On the other side, we have the so-called Riemann surfaces. The simplest example is the sphere.[2]

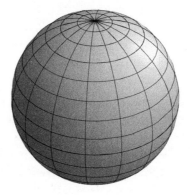

The next example is the torus, the surface in the shape of a donut. I want to emphasize that we are considering here the *surface* of the donut, not its interior.

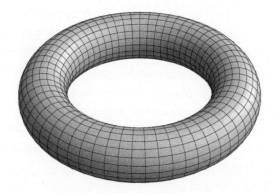

The next example is the surface of a Danish pastry, shown on the next picture (or you can think of it as the surface of a pretzel).

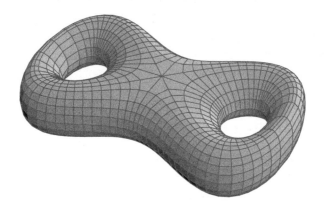

The torus has one "hole," and the Danish has two "holes." There are also surfaces having n holes for any $n = 3, 4, 5,...$ Mathematicians call the number of holes the *genus* of the Riemann surface.* They are named after the German mathematician Bernhard Riemann, who lived in the nineteenth century. His work opened up several important directions in mathematics. Riemann's theory of curved spaces, which we now call Riemannian geometry, is the cornerstone of Einstein's general relativity theory. Einstein's equations describe the force of gravity in terms of the so-called Riemann tensor expressing the curvature of space-time.

At first glance, number theory has nothing in common with the Riemann surfaces. However, it turns out that there are many analogies between them. The key point is that there is another class of objects between these two.

To see this, we have to realize that a Riemann surface may be described by an algebraic equation. For example, consider again a cubic equation such as

$$y^2 + y = x^3 - x^2.$$

As we noted earlier, when we talk about solutions of such an equation, it is important to specify to what numerical system they belong. There are many choices, and different choices give rise to different mathematical theories.

---

*My editor tells me that the pretzels at the German bar near his house are genus–3 (and delicious).

In the previous chapter, we discussed solutions modulo prime numbers, and that's one theory. But we can also look for solutions in *complex numbers*. That's another theory, which yields Riemann surfaces.

People often ascribe almost mystical qualities to complex numbers, as if these are some incredibly complicated objects. The truth is that they are no more complicated than the numbers we discussed in the previous chapter when trying to make sense of the square root of 2.

Let me explain. In the previous chapter, we adjoined to the rational numbers two solutions of the equation $x^2 = 2$, which we denoted by $\sqrt{2}$ and $-\sqrt{2}$. Now, instead of looking at the equation $x^2 = 2$, we look at the equation $x^2 = -1$. Does it look much more complicated than the previous one? No. It has no solutions among rational numbers, but we are not afraid of this. Let's adjoin the two solutions of this equation to the rational numbers. Denote them by $\sqrt{-1}$ and $-\sqrt{-1}$. They solve the equation $x^2 = -1$, that is,

$$\sqrt{-1}^2 = -1, \qquad (-\sqrt{-1})^2 = -1.$$

There is only a minor difference with the previous case. The number $\sqrt{2}$ is not rational, but it is a *real* number, so by adjoining it to the rational numbers, we don't leave the realm of real numbers.

We can think of real numbers geometrically as follows. Draw a line and mark two points on it, which will represent numbers 0 and 1. Then mark the point to the right of 1 whose distance to 1 is equal to the distance between 0 and 1. This point will represent number 2. We represent all other integers in a similar fashion. Next, we mark rational numbers by subdividing the intervals between the points representing the integers. For example, the number $\frac{1}{2}$ is exactly halfway between 0 and 1; the number $\frac{7}{3}$ is one-third of the way from 2 to 3, and so on. Now, the real numbers are, intuitively, in one-to-one correspondence with all points of this line.[3]

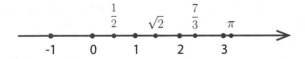

Recall that we encounter number $\sqrt{2}$ as the length of the hypotenuse of the right triangle with legs of length 1. So we mark $\sqrt{2}$ on the line of real numbers by finding a point to the right of 0 whose distance to 0 is equal to the length of this hypotenuse. Likewise, we can mark[4] on this line the number π, which is the circumference of a circle of diameter 1.

On the other hand, the equation $x^2 = -1$ has no solutions among rational numbers, and it also has no solutions among real numbers. Indeed, the square of any real number must be positive or 0, so it cannot be equal to -1. So unlike $\sqrt{2}$ and $-\sqrt{2}$, the numbers $\sqrt{-1}$ and $-\sqrt{-1}$ are not real numbers. But so what? We follow the same procedure and introduce them in exactly the same way as we introduced the numbers $\sqrt{2}$ and $-\sqrt{2}$. And we use the same rules to do arithmetic with these new numbers.

Let's recall how we argued before: we noticed that the equation $x^2 = 2$ had no solutions among the rational numbers. So we created two solutions of this equation, called them $\sqrt{2}$ and $-\sqrt{2}$, and adjoined them to the rational numbers, creating a new numerical system (which we then called a number field). Likewise, now we take the equation $x^2 = -1$ and notice that it also has no solutions among rational numbers. So we *create* two solutions of this equation, denote them by $\sqrt{-1}$ and $-\sqrt{-1}$, and adjoin them to the rational numbers. It's exactly the same procedure! Why should we think of this new numerical system as anything more complicated than our old numerical system, the one with $\sqrt{2}$?

The reason is purely psychological: whereas we can represent $\sqrt{2}$ as the length of a side of a right triangle, we don't have such an obvious geometric representation of $\sqrt{-1}$. But we can manipulate $\sqrt{-1}$ algebraically as effectively as $\sqrt{2}$.

Elements of the new numerical system we obtain by adjoining $\sqrt{-1}$ to the rational numbers are called complex numbers. Each of them may be written as follows:

$$r + s\sqrt{-1},$$

where r and s are rational numbers. Compare this to the formula on p. 73 expressing general elements of the numerical system obtained by adjoining $\sqrt{2}$. We can add any two numbers of this form by adding separately their r-parts and s-parts. We can also multiply any two such numbers by opening the brackets and using the fact that $\sqrt{-1} \cdot \sqrt{-1} = -1$. In a similar way, we can also subtract and divide these numbers.

Finally, we extend the definition of complex numbers by allowing r and s in the above formula to be arbitrary real numbers (not just the rational numbers). Then we obtain the most general complex numbers. Note that it is customary to denote $\sqrt{-1}$ by i (for "imaginary"), but I chose not to do this to emphasize the algebraic meaning of this number: it really is just a square root of -1, nothing more and nothing less. It is just as concrete as the square root of 2. There is nothing mysterious about it.

We can get a feel for how concrete these numbers are by representing them geometrically. Just as the real numbers may be represented geometrically as points of a line, complex numbers may be represented as points of a plane. Namely, we represent the complex number $r + s\sqrt{-1}$ as a point on the plane with coordinates r and s:[5]

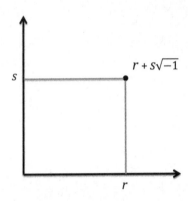

Let's go back to our cubic equation

$$y^2 + y = x^3 - x^2$$

and let's look for solutions x and y that are complex numbers.

A remarkable fact is that the set of all such solutions turns out to be exactly the set of points of a torus depicted earlier. In other words, each point of the torus can be assigned to one and only one pair of complex numbers x, y solving the above cubic equation, and vice versa.[6]

If you've never thought about complex numbers before, your head might be starting to hurt just about now. This is completely natural. Wrapping one's mind around a single complex number is challenging enough, let alone pairs of complex numbers solving some equation. It's not obvious at all that these pairs are in one-to-one correspondence with points on the surface of a donut, so don't be alarmed if you don't see why this is so. In fact, many professional mathematicians would be hard-pressed to prove this surprising and non-trivial result.[7]

In order to convince ourselves that solutions of algebraic equations give rise to geometric shapes, let's look at a simpler situation: solutions over the real numbers instead of complex numbers. For example, consider the equation

$$x^2 + y^2 = 1$$

and let's mark its solutions as points on the plane with coordinates x and y. The set of all such solutions is a circle of radius one, centered at the origin. Likewise, solutions of any other algebraic equation in two *real*-valued variables x and y form a curve on this plane.[8]

Now, complex numbers are in some sense doubles of real numbers (indeed, each complex number is determined by a pair of real numbers), so it's not surprising that solutions of an algebraic equation in *complex*-valued variables x and y form a Riemann surface. (A curve is one-dimensional, and a Riemann surface is two-dimensional, in the sense explained in Chapter 10.)

In addition to real and complex solutions, we may also look for solutions x, y of these equations that take values in a finite field $\{0, 1, 2, ..., p-2, p-1\}$, where p is a prime number. This means that when we substitute x, y in the above cubic equation, say, the left- and right-hand sides become integers that are equal to each other up to an integer multiple of p. This gives us an object that mathematicians call a "curve over a finite field." Of course, these are not really curves. The terminology is due to the fact that when we look for solutions in real numbers, we obtain curves on the plane.[9]

A deep insight of Weil was that the most fundamental object here is an algebraic equation, like the cubic one above. Depending on the choice of the domain where we look for solutions, the same equation gives rise to a surface, a curve, or a bunch of points. But those are nothing but *avatars* of an ineffable being, which is the equation itself, the way Vishnu has ten avatars, or incarnations, in Hinduism. Somewhat serendipitously, in the letter to his sister, André Weil invoked the *Bhagavad-Gita*,[10] a sacred text of Hinduism, in which the doctrine of avatars of Vishnu is believed to appear for the first time.[11] Weil wrote poetically about what happens when the inkling of an analogy between two theories is turned into concrete knowledge:[12]

> Gone are the two theories, gone their troubles and delicious reflections in one another, their furtive caresses, their inexplicable quarrels; alas, we have but one theory, whose majestic beauty can no longer excite us. Nothing is more fertile than these illicit liaisons; nothing gives more pleasure to the connoisseur.... The pleasure comes from the illusion and the kindling of the senses; once the illusion disappears and knowledge is acquired, we attain indifference; in the *Gita* there are some lucid verses to that effect. But let's go back to algebraic functions.

The connection between Riemann surfaces and curves over finite fields should now be clear: both come from the same kind of equations, but we look for solutions in different domains, either finite fields or complex numbers. On the other hand, "any argument or result in number theory can be translated, word for word," to curves over finite fields, as Weil put it in his letter.[13] Weil's idea was therefore that curves over finite fields are the objects that intermediate between number theory and Riemann surfaces.

Thus, we find a bridge, or a "turntable" – as Weil called it – between number theory and Riemann surfaces, and that is the theory of algebraic curves over finite fields. In other words, we have three parallel tracks, or columns:

Number Theory   *Curves over Finite Fields*   *Riemann Surfaces*

Weil wanted to exploit this in the following way: take a statement in one of the three columns and translate it into statements in the other columns. He wrote to his sister:[14]

> My work consists in deciphering a trilingual text; of each of the three columns
> I have only disparate fragments; I have some ideas about each of the three lan-
> guages: but I know as well there are great differences in meaning from one column
> to another, for which nothing has prepared me in advance. In the several years I
> have worked at it, I have found little pieces of the dictionary.

Weil went on to find one of the most spectacular applications of his Rosetta stone: what we now call the Weil conjectures. The proof of these conjectures[15] greatly stimulated the development of mathematics in the second half of the twentieth century.

Let's go back to the Langlands Program. Langlands' original ideas concerned the left column of Weil's Rosetta stone; that is, number theory. Langlands related representations of the Galois groups of number fields, which are objects studied in number theory, to automorphic functions, which are objects in harmonic analysis – an area of mathematics that is far removed from number theory (and also far away from other columns of the Rosetta stone). Now we can ask whether this kind of relation may also be found if we replace the Galois groups by some objects in the the middle and the right columns of Weil's Rosetta stone.

It is fairly straightforward to translate Langlands' relation to the middle column because all the necessary ingredients are readily available. Galois groups of number fields should be replaced here by the Galois groups relevant to curves over finite fields. There also exists a branch of harmonic analysis that studies suitable automorphic functions. Already in his original work, Langlands related representations of the Galois groups and automorphic functions relevant to the middle column.

However, it is not at all clear how to translate this relation to the right column of the Rosetta stone. In order to do this, we have to find geometric analogues of the Galois groups and automorphic functions in the theory of Riemann surfaces. When Langlands first formulated his ideas, the former was known, but the latter was a big mystery. It wasn't until the 1980s that the appropriate notion was found, starting with the pioneering work by a brilliant Russian mathematician Vladimir Drinfeld. This enabled the translation of the Langlands relation to the third column of the Rosetta stone.

Let's discuss first the geometric analogue of the Galois group. It is the so-called *fundamental group* of a Riemann surface.

The fundamental group is one of the most important concepts in the mathematical field of topology, which focuses on the most salient features of geometric shapes (such as the number of "holes" in a Riemann surface).

Consider, for example, a torus. We pick a point on it – call it P – and look at the closed paths starting and ending at this point. Two such paths are shown on the picture.

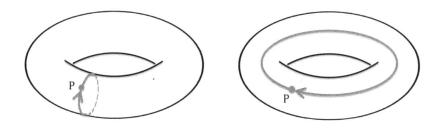

Likewise, the fundamental group of any given Riemann surface consists of such closed paths on this Riemann surface starting and ending at the same fixed point P.[16]

Given two paths starting and ending at the point P, we construct another path as follows: we move along the first path and then move along the second path. This way we obtain a new path, which will also start and end at the point P. It turns out that this "addition" of closed paths satisfies all properties of a group listed in Chapter 2. Thus, we find that these paths indeed form a group.[17]

You may have noticed that the rule of addition of paths in the fundamental group is similar to the rule of addition of braids in the braid groups, as defined in Chapter 5. This is not accidental. As explained in Chapter 5, braids with n threads may be viewed as paths on the space of collections of n distinct points on the plane. In fact, the braid group B_n is precisely the fundamental group of this space.[18]

It turns out that the two paths on the torus shown on the above picture commute with each other; that is, adding them in two possible orders gives us the same element of the fundamental group.[19] The most general element of the fundamental group of the torus is therefore obtained by following the first path M times and then following the second path N times, where M and N are two integers (if M is negative, then we follow the first path $-M$ times in the opposite direction, and similarly for negative N). Since the two basic paths commute with each other, the order in which we follow these paths does not matter; the result will be the same.

For other Riemann surfaces, the structure of the fundamental group is more complicated.[20] Different paths do not necessarily commute with each other. This is similar to braids with more than two threads not commuting with each other, as we discussed in Chapter 5.

It has been known for some time that there is a deep analogy between the Galois groups and the fundamental groups.[21] This provides the answer to our first question: what is the analogue of the Galois group in the right column of Weil's Rosetta stone? It is the fundamental group of the Riemann surface.

Our next question is to find suitable analogues of the automorphic functions, the objects that appear on the other side of the Langlands relation. And here we have to make a quantum leap. The good old functions turn out to be inadequate. They need to be replaced by more sophisticated objects of modern mathematics called *sheaves*, which will be described in Chapter 14.

This was proposed by Vladimir Drinfeld in the 1980s. He gave a new formulation of the Langlands Program that applies to the middle and the right columns, which concern curves over finite fields and Riemann surfaces, respectively. This formulation became known as the geometric Langlands Program.

In particular, Drinfeld found the analogues of the automorphic functions suitable for the right column of Weil's Rosetta stone.

I met Drinfeld at Harvard University in the spring of 1990. Not only did he get me excited about the Langlands Program, he also told me that I had a role to play in its development. That's because Drinfeld saw a connection between the geometric Langlands Program and the work I did as a student in Moscow. The results of this work were essential in Drinfeld's new approach, and this in turn shaped my mathematical life: the Langlands Program has played a dominant role in my research ever since.

So let us return to Moscow, and see where I went after finishing my first paper, on braid groups.

Chapter 10

Being in the Loop

In Moscow in the fall of 1986, I was in the third year of my studies at Kerosinka. With the braid group paper finished and submitted, Fuchs had a question for me: "What do you want to do next?"

I wanted another problem to solve. It turned out that for several years Fuchs had been working with his former student Boris Feigin on representations of "Lie algebras." Fuchs said it was an active area with many unsolved problems and with close ties to quantum physics.

That sure caught my attention. Even though Evgeny Evgenievich had "converted" me to math, and even though I was enchanted by mathematics, I had never lost my childhood fascination with physics. That the worlds of math and quantum physics might come together was exciting for me.

Fuchs handed me an eighty-page research paper he and Feigin had written.

"I first thought of giving you a textbook on Lie algebras," he said. "But then I thought, why not just give you this paper?"

I put the paper carefully in my backpack. It was still unpublished at the time, and, thanks to the tight controls that Soviet authorities (afraid that people would make copies of banned literature, like books of Solzhenitsyn or *Doctor Zhivago*) placed on photocopiers, there were only a handful of copies available in the entire world. Very few people had ever gotten to see this paper – Feigin later joked that I may have been the only one who had read it from beginning to end.

It was written in English and was supposed to appear in a collection of papers published in the U.S. But the publisher badly mismanaged the book, and its publication was delayed for some fifteen years. By then, most of the results were reproduced elsewhere, so it wasn't much read after it came out either. Nevertheless, the article became famous, and Feigin and Fuchs eventually got their due credit. Their paper has been widely cited in the literature (as a "Moscow Preprint"), and even a new term was coined, "Feigin–Fuchs representations," to refer to the new representations of Lie algebras they studied in this paper.

As I started reading the paper, my first question was: what are these objects that carry such a strange name, "Lie algebras"? The paper that Fuchs gave me assumed quite a bit of knowledge about topics I'd never studied, so I went to a bookstore and bought all the textbooks on Lie algebras I could find. Whatever I could not find, I borrowed from the library at Kerosinka. I was reading all these books in parallel with the Feigin–Fuchs article. This experience shaped my learning style. Since then, I've never been satisfied with one source; I try to find all available sources and devour them.

To explain what Lie algebras are, I first need to tell you about "Lie groups." Both are named after a Norwegian mathematician Sophus Lie (pronounced LEE) who invented them.

Mathematical concepts populate the Kingdom of Mathematics, just like species of animals populating the Animal Kingdom: they are linked to each other, form families and subfamilies, and often two different concepts mate and produce an offspring.

The concept of a group is a good example. Think of groups as analogues of birds, which form a class in the Animal Kingdom, or Animalia (called class Aves). That class is split into twenty-three orders; each order in turn splits into families, and each of those splits into genera. For example, the African fish eagle belongs to the order of Accipitriformes, the family of Accipitridae, and the genus of *Haliaeetus* (compared to these names, "Lie group" doesn't sound so exotic!). Likewise, groups form a large class of mathematical concepts, and within this class there are different "orders," "families," and "genera."

For example, there is an order of finite groups that includes all groups with finitely many elements. The group of symmetries of a square table, which we discussed in Chapter 2, consists of four elements, so it is a finite group. Likewise, the Galois group of a number field obtained by adjoining the solutions of a polynomial equation to the rational numbers is a finite group (for example, in the case of a quadratic equation it has two elements). The class of finite groups is further subdivided into families, such as the family of Galois groups. Another family consists of the crystallographic groups, which are the groups of symmetries of various crystals.

There is also another order, of infinite groups. For example, the group of integers is infinite, and so is the braid group B_n, which we discussed in Chapter 5, for each fixed $n = 2, 3, 4,...$ (B_n consists of braids with n threads; there are infinitely many such braids). The group of rotations of a round table, which consists of all points on a circle, is also an infinite group.

But there is an important difference between the group of integers and the circle group. The group of integers is discrete; that is to say, its elements do not combine into a continuous geometric shape in any natural sense. We can't move continuously from one integer to the next; we jump from one to another. In contrast, we can change the angle of rotation continuously between 0 and 360 degrees. And together, these angles combine into a geometric shape: namely, the circle. Mathematicians call such shapes *manifolds*.

The group of integers and the braid groups belong to the family of discrete infinite groups in the Kingdom of Mathematics. And the circle group belongs to another family, that of Lie groups. Put simply, a Lie group is a group whose elements are points of a manifold. So this concept is the offspring resulting from the marriage of two mathematical concepts: group and manifold.

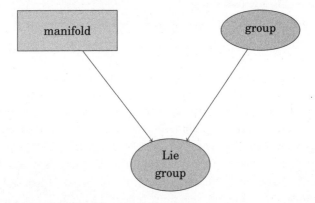

Here is the tree of group-related concepts that we will discuss in this chapter (some of these concepts have not yet been introduced, but will be later in the chapter).

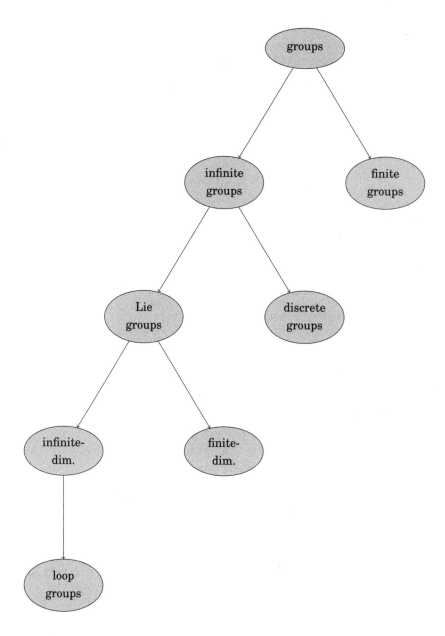

Many symmetries arising in nature are described by Lie groups, and that's why they are so important to study. For example, the group $SU(3)$ we talked about in Chapter 2, which is used to classify elementary particles, is a Lie group.

Here is another example of a Lie group: the group of rotations of a sphere. A rotation of a round table is determined by its angle. But in the case of a sphere, there is more freedom: we have to specify the axis as well as the angle of rotation, as shown on the picture. The axis can be any line passing through the center of the sphere.

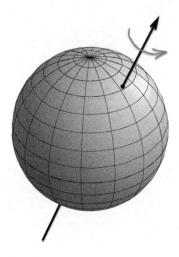

The group of rotations of the sphere has a name in math: the special orthogonal group of the 3-dimensional space, or, as it is commonly abbreviated, $SO(3)$. We can think of the symmetries of the sphere as transformations of the 3-dimensional space in which the sphere is embedded. These transformations are orthogonal, meaning that they preserve all distances.[1] Incidentally, this gives us a 3-dimensional representation of the group $SO(3)$, a concept we introduced in Chapter 2.

Likewise, the group of rotations of the round table, which we have discussed above, is called $SO(2)$; these rotations are special orthogonal transformations of the plane, which is 2-dimensional. Thus, we have a 2-dimensional representation of the group $SO(2)$.

The groups $SO(2)$ and $SO(3)$ are not only groups but also manifolds (that is, geometric shapes). The group $SO(2)$ is the circle, which is a manifold. So $SO(2)$ is a group and a manifold. That's why we say that it is a Lie group. Likewise, elements of the group $SO(3)$ are points of another manifold, but it is more tricky to visualize it. (Note that this manifold is *not* a sphere.) Recall that each rotation of the sphere is determined by the axis and the angle of rotation. Now observe that each point of the sphere gives rise to an axis of rotation: the line connecting this point and the center of the sphere. And the angle of rotation is the same as a point of a circle. So an element of the group $SO(3)$ is determined by a point of the sphere (it defines the axis of rotation) together with a point of a circle (it defines the angle of rotation).

Perhaps we should start with a simpler question: what is the dimension of $SO(3)$? To answer this question, we need to discuss the meaning of dimension more systematically. We have already mentioned in Chapter 2 that the world around us is three-dimensional. That is to say, in order to specify a position of a point in space, we need to specify three numbers, or coordinates, (x, y, z). A plane, on the other hand, is two-dimensional: a position on the plane is specified by two coordinates, (x, y). And a line is one-dimensional: there is only one coordinate, x.

But what is the dimension of a circle? It is tempting to say that the circle is two-dimensional because we usually draw the circle on a plane, which is two-dimensional. Each point of the circle, when viewed as a point of the plane, is described by two coordinates. But the mathematical definition of the dimension of a given geometric object (such as a circle) is the number of independent coordinates we need *on this object* to pinpoint any location on it. This number has nothing to do with the dimension of the landscape into which the object is embedded (such as a plane). Indeed, a circle can also be embedded into a three-dimensional space (think of a ring on one's finger), or a space of even larger dimension. What matters is that for a particular circle, the position of any point on it can be described by one number, namely, the angle. This is the sole coordinate on the circle. That's why we say that the circle is one-dimensional.

Of course, in order to speak of the angle, we need to pick a reference point on the circle corresponding to angle 0. Likewise, in order to assign a coordinate x to each point on the line we need to pick a reference point on it corresponding to $x = 0$. We can set up a coordinate system on a given object in many different ways. But each of these coordinate systems will have the same *number* of coordinates, and it's this number that is called the *dimension* of this object.

Note that as we zoom in and look at a smaller and smaller neighborhood of a point of the circle, the curvature of the circle all but disappears. There is practically no difference between a small neighborhood of a point on the circle and a small neighborhood of the same point on the tangent line to the circle; this is the line that is the closest approximation to the circle near this point.

This shows that the circle and the line have the same dimension.[2]

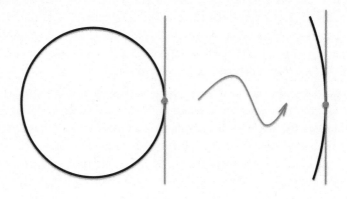

As we zoom in on a point, the circle and the tangent line appear
closer and closer to each other.

Likewise, the sphere is embedded into a three-dimensional space, but its intrinsic dimension is two. Indeed, there are two independent coordinates on the sphere: latitude and longitude. We know them well because we use them to determine the position on the surface of the Earth, which is close to the shape of a sphere. The mesh on the sphere that we see on the above picture is made of the "parallels" and "meridians," which correspond to fixed values of latitude and longitude. The fact that there are two coordinates on the sphere tells us that it is two-dimensional.

What about the Lie group $SO(3)$? Every point of $SO(3)$ is a rotation of the sphere, so we have three coordinates: the axis of rotation (which may be specified by a point at which the axis pierces the sphere) is described by two coordinates, and the angle of rotation gives rise to the third coordinate. Hence the dimension of the group $SO(3)$ is equal to three.

Thinking about a Lie group, or any manifold, of more than three dimensions can be very challenging. Our brain is wired in such a way that we can only imagine geometric shapes, or manifolds, in dimensions up to three. Even imagining the four-dimensional combination of space and time is a strenuous task: we just don't perceive the time (which constitutes the fourth dimension) as an equivalent of a spatial dimension. What about higher dimensions? How can we analyze five- or six- or hundred-dimensional manifolds?

Think about this in terms of the following analogy: works of art give us two-dimensional renderings of three-dimensional objects. Artists paint two-dimensional projections of those objects on the canvas and use the technique of perspective to create the illusion of depth, the third dimension, in their paintings. Likewise, we can imagine four-dimensional objects by analyzing their three-dimensional projections.

Another, more efficient way to imagine a fourth dimension is to think of a four-dimensional object as a collection of its three-dimensional "slices." This would be similar to slicing a loaf of bread, which is three-dimensional, into slices so thin that we could think of them as being two-dimensional.

If the fourth dimension represents time, then this four-dimensional "slicing" is known as photography. Indeed, snapping a picture of a moving person gives us a three-dimensional slice of a four-dimensional object representing that person in the four-dimensional space-time (this slice is then projected onto a plane). Taking several pictures in succession, we obtain a collection of such slices. If we run these pictures quickly in front of our eyes, we can see that movement. This is of course the basic idea of cinema.

We can also convey the impression of the person's movement by juxtaposing the pictures. At the beginning of the twentieth century, artists got interested in this idea and used it as a way to include the fourth dimension into their paintings, to render them dynamic. A milestone in this direction was Marcel Duchamp's 1912 painting *Nude Descending a Staircase, No. 2*.

It is interesting to note that Einstein's relativity theory, which demonstrated that space and time cannot be separated from each other, appeared around the same time. This brought the notion of the four-dimensional space-time continuum to the forefront of physics. In parallel, mathematicians such as Henri Poincaré were delving deeper into the mysteries of higher-dimensional geometry and transcending the Euclidean paradigm.

Duchamp was fascinated with the idea of the fourth dimension as well as non-Euclidean geometry. Reading E.P. Jouffret's book *Elementary Treatise on Four-Dimensional Geometry and Introduction to the Geometry of n Dimensions*, which in particular presented the groundbreaking ideas of Poincaré, Duchamp left the following note:[3]

> The shadow cast by a 4-dimensional figure on our space is a 3-dimensional shadow (see Jouffret – Geom. of 4-dim., page 186, last 3 lines).... by analogy with the method by which architects depict a plan of each story of a house, a 4-dimensional figure can be represented (in each one of its stories) by three-dimensional sections. These different stories will be bound to one another by the 4th dim.

According to art historian Linda Dalrymple Henderson,[4] "Duchamp found something deliciously subversive about the new geometries with their challenge to so many long-standing 'truths.'" The interest of Duchamp and other artists of that era in the fourth dimension, she writes, was one of the elements that led to the birth of abstract art.

Thus, mathematics informed art; it allowed artists to see hidden dimensions and inspired them to expose, in a tantalizing aesthetic form, some profound truths about our world. The works of modern art they created helped elevate our perception of reality, affecting our collective consciousness. This in turn influenced the next generations of mathematicians. Philosopher of science Gerald Holton put this eloquently:[5]

> Indeed, a culture is kept alive by the interaction of all its parts. Its progress is an alchemical process, in which all its varied ingredients can combine to form new jewels. On this point, I imagine that Poincaré and Duchamp are in agreement with me and with each other, both having by now undoubtedly met somewhere in that hyperspace which, in their different ways, they loved so well.

Mathematics enables us to perceive geometry in all of its incarnations, shapes, and forms. It is a universal language that applies equally well in all dimensions, whether we can visualize the corresponding objects or not, and allows us to go far beyond our limited visual imagination. In fact, Charles Darwin wrote that mathematics endows us with "an extra sense."[6]

For example, though we cannot imagine a four-dimensional space, we can actualize it mathematically. We simply represent points of this space as quadruples of numbers (x, y, z, t), just like we represent points of the three-dimensional space by triples of numbers (x, y, z). In the same way, we can view points of an n-dimensional flat space, for any natural number n, as n-tuples of numbers (we can analyze these in the same way as the rows of a spreadsheet, as we discussed in Chapter 2).

Perhaps I need to explain why I refer to these spaces as being flat. A line is clearly flat and so is a plane. But it's not as obvious that we should think of the three-dimensional space as flat. (Note that I am not talking here about various curved manifolds embedded into the three-dimensional space, such as a sphere or a torus. I am talking about the three-dimensional space itself.) The reason is that it has no curvature. The precise mathematical definition of curvature is subtle (it was given by Bernhard Riemann, the creator of Riemann surfaces), and we won't go into the details now as this is tangential to our immediate goals. A good way to think about the flatness of the three-dimensional space is to realize that it has three infinite coordinate axes that are perpendicular to each other, just as a plane has two perpendicular coordinate axes. Likewise, an n-dimensional space, with n perpendicular coordinate axes, has no curvature and hence is flat.

Physicists have thought for centuries that we inhabit a flat three-dimensional space, but, as we discussed in the Preface, Einstein has shown in his general relativity theory that gravity causes space to curve (the curvature is small, so that we don't notice it in our everyday life, but it is non-zero). Therefore our space is in fact an example of a curved three-dimensional manifold.

This brings up the question of how a curved space could possibly exist by itself, without being embedded into a flat space of higher dimension, the way a sphere is embedded into a flat three-dimensional space. We are used to thinking that the space we live in is flat, and so in our everyday experience curved shapes seem to appear only within the confines of that flat space. But this is a misunderstanding, an artifact of our narrow perception of reality. And the irony is that the space we live in isn't flat to begin with! Mathematics gives us a way out of this trap: as Riemann showed, curved spaces do exist intrinsically, as objects of their own making, without a flat space containing them. What we need to define such a space is a rule of measuring distances between any two points of this space (this rule must satisfy certain natural properties); this is what mathematicians call a *metric*. The mathematical concepts of metric and the curvature tensor, introduced by Riemann, are the cornerstones of Einstein's general relativity theory.[7]

Curved shapes, or manifolds, can have arbitrarily high dimensions. Recall that the circle is defined as the set of points on a plane equidistant from a given point (or, as my examiner at MGU insisted, the set of *all* such points!). Likewise, a sphere is the set of all points in the three-dimensional space equidistant from a given point. Now, define a higher-dimensional analogue of a sphere – some call it a hypersphere – as the set of points equidistant from a given point in the n-dimensional space. This condition gives us one constraint on the n coordinates. Therefore the dimension of the hypersphere inside the n-dimensional space is $(n-1)$. Further, we can study the Lie group of rotations of this hypersphere.[8] It is denoted by $SO(n)$.

From the point of view of the taxonomy of groups in the Kingdom of Mathematics, the family of Lie groups is subdivided into two genera: that of finite-dimensional Lie groups (such as the circle group and the group $SO(3)$) and that of infinite-dimensional Lie groups. Note that any finite-dimensional Lie group is already infinite, in the sense that it has infinitely many elements. For example, the circle group has infinitely many elements (these are the points of the circle). But it is one-dimensional because all of its elements may be described by one coordinate (the angle). For an infinite-dimensional Lie group, we need infinitely many coordinates in order to describe its elements. This kind of "double infinity" is really hard to imagine. Yet, such groups do arise in nature, so we need to study them as well. I will now describe an example of an infinite-dimensional Lie group known as a *loop group*.

To explain what it is, let's first consider loops in the three-dimensional space. Simply put, a loop is a closed curve, such as the one shown on the left-hand side of the picture below. We have already seen them when we talked about braid groups (we called them "knots").[9] I want to stress that non-closed curves, such as the one shown on the right-hand side of the picture, are *not* considered as loops.

Similarly, we can also consider loops (that is, closed curves) inside any manifold M. The space of all these loops is called the loop space of M.

As we will discuss in more detail in Chapter 17, these loops play a big role in string theory. In conventional quantum physics, the fundamental objects are elementary particles, such as electrons or quarks. They are point-like objects,

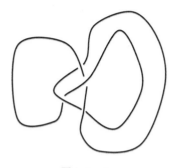
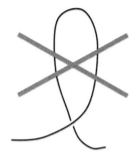

This is a loop   This is *not* a loop

with no internal structure; that is, zero-dimensional. In string theory it is postulated that fundamental objects of nature are one-dimensional strings.[10] A closed string is nothing but a loop embedded in a manifold M (the space-time). That's why loop spaces are the bread and butter of string theory.

Now let's consider the loop space of the Lie group $SO(3)$. Its elements are loops in $SO(3)$. Let's look at one of these loops closely. First of all, it is similar to the loop pictured above. Indeed, $SO(3)$ is three-dimensional, so on a small scale it looks like the three-dimensional flat space. Second, each point on this loop is an element of $SO(3)$, that is, a rotation of the sphere. Hence our loop is a sophisticated object: it is a one-parameter collection of rotations of the sphere. Given two such loops, we can produce a third by composing the corresponding rotations of the sphere. Thus, the loop space of $SO(3)$ becomes a group. We call it the loop group of $SO(3)$.[11] It's a good example of an infinite-dimensional Lie group: we really cannot describe its elements by using a finite number of coordinates.[12]

The loop group of any other Lie group (for example, the group $SO(n)$ of rotations of a hypersphere) is also an infinite-dimensional Lie group. These loop groups arise as symmetry groups in string theory.

The second concept relevant to the paper by Feigin and Fuchs that I was studying was the concept of a Lie algebra. Each Lie algebra is in some sense a simplified version of a given Lie group.

The term "Lie algebra" is bound to create some confusion. When we hear the word "algebra," we think of the stuff we studied in high school, such as solving quadratic equations. However, now the word "algebra" is used in a different connotation: as part of the indivisible term "Lie algebra" referring to mathematical objects with specific properties. Despite what the name suggests, these objects do not form a family in the class of all algebras, the way Lie groups form a family in the class of all groups. But nevermind, we'll just have to live with this inconsistency of terminology.

To explain what a Lie algebra is, I first have to tell you about the concept of the *tangent space*. Don't worry, we are not going off on a tangent; we follow one of the key ideas of calculus called "linearization," that is, approximation of curved shapes by linear, or flat, ones.

For example, the tangent space to a circle at a given point is the line that passes through this point and is the line closest to the circle among all lines passing through this point. We have already encountered it above when we talked about the dimension of the circle. The tangent line just touches the circle at this particular point, ever so slightly, whereas all other lines passing through this point cross the circle at another point as well, as shown on the picture.

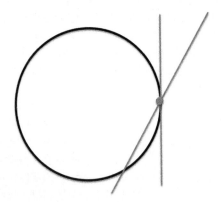

Likewise, any curve (that is, a one-dimensional manifold) can be approximated near a given point by a tangent line. René Descartes, who described an efficient method for computing these tangent lines in his *Géométrie*, published in 1637, wrote:[13] "I dare say that this is not only the most useful and most general problem in geometry that I know, but even that I have ever desired to know." Similarly, a sphere can be approximated at a given point by a tangent plane. Think of a basketball: when we put it on the floor, it touches the floor at one point, and the floor becomes its tangent plane at that point.[14]

And an n-dimensional manifold may be approximated at a given point by a flat n-dimensional space.

Now, on any Lie group we have a special point, which is the identity element of this group. We take the tangent space to the Lie group at this point – and *voilà*, that's the Lie algebra of this Lie group. So each Lie group has its own Lie algebra, which is like a younger sister of the Lie group.[15]

For example, the circle group is a Lie group, and the identity element of this group is a particular point on this circle[16] corresponding to the angle 0. The tangent line at this point is therefore the Lie algebra of the circle group. Alas, we cannot draw a picture of the group $SO(3)$ and its tangent space because they are both three-dimensional. But the mathematical theory describing tangent spaces is set up in such a way that it works equally well in all dimensions. If we want to imagine how things work, we can model them on one- or two-dimensional examples (like a circle or a sphere). In doing so, we use lower-dimensional manifolds as metaphors for more complicated, higher-dimensional manifolds. But we don't have to do this; the language of mathematics enables us to transcend our limited visual intuition. Mathematically, the Lie algebra of an n-dimensional Lie group is an n-dimensional flat space, also known as a vector space.[17]

There is more. The operation of multiplication on a Lie group gives rise to an operation on its Lie algebra: given any two elements of the Lie algebra, we can construct a third. The properties of this operation are more difficult to describe than the properties of multiplication in a Lie group, and they are not essential to us at the moment.[18] An example, which would be familiar to those readers who have studied vector calculus, is the operation of *cross-product* in the three-dimensional space.[19] If you have seen this operation, you may have wondered about its weird-looking properties. And guess what, this operation actually makes the three-dimensional space into a Lie algebra!

It turns out that this is in fact the Lie algebra of the Lie group $SO(3)$. So the esoteric-looking operation of cross-product is inherited from the rule of composition of rotations of the sphere.

You may be wondering why we care about Lie algebras if the operation on them is so weird-looking. Why not stick with Lie groups? The main reason is that, unlike a Lie group, which is usually curved (like a circle), a Lie algebra is a flat space (like a line, a plane, and so on). This makes the study of Lie algebras much simpler than the study of Lie groups.

For example, we can talk about the Lie algebras of loop groups.[20] These Lie algebras, which we should think of as simplified versions of the loop groups, are called *Kac–Moody algebras*, after two mathematicians: Victor Kac (Russian-born, emigrated to the U.S., now Professor at MIT) and Robert Moody (British-born, emigrated to Canada, now Professor at University of Alberta). They started investigating these Lie algebras independently in 1968. Since then, the theory of Kac–Moody algebras has been one of the hottest and fastest growing areas of mathematics.[21]

It was these Kac–Moody algebras that Fuchs had suggested as the topic of my next research project. When I started learning all this, I saw that I had to do a lot of studying before I could get to the point where I could do something on my own. But I was fascinated with the subject.

Fuchs lived in the northeastern part of Moscow, not far from a train station where I could catch a train to my hometown. I used to go home every Friday for the weekend, so Fuchs suggested that I come to his place every Friday at 5 pm and then take the train home after our meeting. I would usually work with him for about three hours (during which he would also feed me dinner), and then I would catch the last train, arriving home around midnight. Those meetings played a big role in my mathematical education. We had them week after week, the entire fall semester of 1986, and then the spring semester of 1987 as well.

It wasn't until January 1987 that I finished reading the long paper by Feigin and Fuchs and felt that I could start working on my research project. By that time, I was able to get a pass to the Moscow Science Library, a huge repository of books and journals, not only in Russian (many of which Kerosinka's library also had), but in other languages as well. I started going there regularly to pore over dozens of math journals, looking for articles about Kac–Moody algebras and related subjects.

I was also eager to learn about their applications to quantum physics, which was of course a huge draw for me. As I mentioned above, Kac–Moody algebras play an important role in string theory, but they also appear as symmetries of models of two-dimensional quantum physics. We live in a three-dimensional space, so realistic models describing our world should be three-dimensional. If we include time, we get four dimensions. But mathematically, nothing precludes us from building and analyzing models describing worlds of other dimensions. The models in dimensions less than three are simpler, and we have a better chance of solving them. We can then use what we learn to tackle the more sophisticated three- and four-dimensional models.

This is in fact one of the main ideas of the subject called "mathematical physics" – study models of different dimensions that may not be directly applicable to our physical world, but share some of the salient features of the realistic models.

Some of these low-dimensional models also have real-world applications. For example, a very thin metal layer may be viewed as a two-dimensional system and hence may be effectively described by a two-dimensional model. A famous example is the so-called Ising model of interacting particles occupying the nodes of a two-dimensional lattice. The exact solution of the Ising model by Lars Onsager provided valuable insights into the phenomenon of spontaneous magnetization, or ferromagnetism. At the core of Onsager's calculation was a hidden symmetry of this model, underscoring once again symmetry's paramount role in understanding physical systems. It was subsequently understood that this symmetry is described by the so-called Virasoro algebra, a close cousin of Kac–Moody algebras.[22] (In fact, it was the Virasoro algebra that was the main subject of the paper by Feigin and Fuchs which I was studying.) There is also a large class of models of this type in which symmetries are described by the Kac–Moody algebras proper. The mathematical theory of Kac–Moody algebras is essential for understanding these models.[23]

Kerosinka's library subscribed to a publication called *Referativny Zhurnal*, the Journal of References. This journal, published monthly, had short reviews of all new articles, in all languages, organized by subject, with a short summary of each. I started reading it regularly, and what a valuable source it turned out to be! Every month a new volume about math papers would come, and I would fish through the relevant sections trying to find something of interest. If I found something that sounded exciting, I would write down the reference and get it on my next visit to the Moscow Science Library. This way, I discovered a lot of interesting stuff.

One day, while turning the pages of the *Referativny Zhurnal*, I stumbled upon a review of a paper by a Japanese mathematician Minoru Wakimoto, which was published in one of the journals I was paying close attention to, *Communications in Mathematical Physics*. The review did not say much, but the title referred to the Kac–Moody algebra associated to the group of rotations of the sphere, $SO(3)$, so I took down the reference and on my next visit to the Science Library I read the article.

In it, the author constructed novel realizations of the Kac–Moody algebra associated to $SO(3)$. To give the gist of what they are, I will use the language of quantum physics (which is relevant here because Kac–Moody algebras describe symmetries of models of quantum physics). Realistic quantum models, like those describing the interaction of elementary particles, are quite complicated. But we can construct much simpler, idealized, "free-field models," in which there is no or almost no interaction. The quantum fields in these models are "free" from each other, hence the name.[24] It is often possible to realize a complicated, and hence more interesting, quantum model inside one of these free-field models. This allows us to dissect and deconstruct the complicated models, and perform computations that are not accessible otherwise. Such realizations are very useful as the result. However, for quantum models with Kac–Moody algebras as symmetries, the known examples of such free-field realizations had been rather narrow in scope.

As I was reading Wakimoto's paper, I saw right away that the result could be interpreted as giving the broadest possible free-field realization in the case of the simplest Kac–Moody algebra, the one associated to $SO(3)$. I understood the importance of this result, and it made me wonder: where did this realization come from? Is there a way to generalize it to other Kac–Moody algebras? I felt that I was ready to tackle these questions.

How to describe the excitement I felt when I saw this beautiful work and realized its potential? I guess it's like when, after a long journey, suddenly a mountain peak comes in full view. You catch your breath, take in its majestic beauty, and all you can say is "Wow!" It's the moment of revelation. You have not yet reached the summit, you don't even know yet what obstacles lie ahead, but its allure is irresistible, and you already imagine yourself at the top. It's yours to conquer now. But do you have the strength and stamina to do it?

Chapter 11

Conquering the Summit

By summer, I was prepared to share my findings with Fuchs. I knew he would be as excited about Wakimoto's paper as I was. I went to see Fuchs at his dacha, but when I arrived, he told me that there was a slight problem: he had made appointments with me and with his collaborator and former student Boris Feigin on the same day – inadvertently, he said, although I didn't believe him (much later Fuchs confirmed that this was indeed intentional).

Fuchs had introduced me to Feigin a few months earlier. It was before one of Gelfand's seminars, soon after I finished my paper on braid groups and was starting to read the article by Feigin and Fuchs. Prompted by Fuchs, I asked Feigin for suggestions as to what else I should be reading. Boris Lvovich, as I addressed him, was then thirty-three years old but already considered one of the biggest stars in the Moscow mathematical community. Wearing a pair of jeans and well-worn sneakers, he appeared to be quite shy. Large thick glasses covered his eyes, and for the most part during our conversation he looked down to avoid making eye contact. Needless to say, I was also shy and not too sure of myself: I was just a beginning student, and he was already a famous mathematician. So this wasn't the most engaging of encounters. But every once in a while he would raise his eyes and look at me with a big disarming smile, and this broke the ice. I could sense his genuine kindness.

However, Feigin's initial suggestion startled me: he told me that I should read the book *Statistical Physics* by Landau and Lifshitz, a prospect that I found absolutely dreadful at the time, partly because of the resemblance, in size and weight, between that thick volume and the textbook on the history of the Communist Party that we all had to study in school.

In Feigin's defense, this was a solid advice – this book is indeed important, and eventually my research turned precisely in that direction (even though I have to admit, to my shame, that I still haven't read the book). But at the time this idea did not resonate with me at all, and perhaps it was partly for this reason that our initial conversation didn't go anywhere. In fact, I never spoke to Feigin again, besides saying "Hello" when I saw him at Gelfand's seminar, until that day at Fuchs' dacha.

Soon after my arrival, I saw Feigin through the window dismounting his bike. After greetings and some small talk, we all sat down at a round table in the kitchen and Fuchs asked me, "So, what's new?"

"Well... I have found this interesting paper by a Japanese mathematician, Wakimoto."

"Hmmm..." Fuchs turned to Feigin: "Do you know about this?"

Feigin shook his head no, and Fuchs said to me, "He always knows everything... But it's good that he hasn't seen this paper – then it will be interesting for him to hear you as well."

I set out to describe Wakimoto's work to both of them. As expected, they were both very interested. This was the first time that I had a chance to discuss mathematical concepts in-depth with Feigin, and instantly I had a feeling that we clicked. He was listening attentively and asking all the right questions. He clearly understood the importance of this stuff, and even though his demeanor remained relaxed and casual, he seemed to be excited about it. Fuchs was mostly looking on, and I am sure he was happy that his secret plan to get Feigin and me more closely acquainted worked so well. It really was an amazing conversation. I felt like I was on the verge of something important.

Fuchs seemed to feel the same way. As I was leaving, he told me, "Well done. I wish this were your paper. But I think you are now ready to take it to the next level."

I went back home and continued to study the questions raised by Wakimoto's paper. Wakimoto didn't give any explanations for his formulas. I was doing what amounted to forensic work – trying to find the traces of a big picture behind his formulas.

A few days later, the picture started to emerge. In a flash of insight, as I was pacing around my dorm room, I realized that Wakimoto's formulas came from geometry. This was a startling discovery because Wakimoto's approach was entirely algebraic – there were no hints of geometry.

To explain my geometric interpretation, let's revisit the Lie group $SO(3)$ of symmetries of the sphere and its loop group. As explained in the previous chapter, an element of the loop group of $SO(3)$ is a collection of elements of $SO(3)$, one element of $SO(3)$ for every point of the loop. Each of those elements of $SO(3)$ acts on the sphere by a particular rotation. This implies that each element of the loop group of $SO(3)$ gives rise to a symmetry of the loop space of the sphere.[1]

I realized that I could use this information to obtain a representation of the Kac–Moody algebra associated to $SO(3)$. This does not yet give us Wakimoto's formulas. To get them, we have to modify the formulas in a certain radical way. Think about it as turning a coat inside out. We can do this to any coat, but in most cases, the garment then becomes unusable – we can't wear it in public. However, there are coats that can be worn on either side. And the same was true of Wakimoto's formulas.

Armed with this new understanding, I immediately tried to generalize Wakimoto's formulas to other, more complicated Kac–Moody algebras. The first, geometric, step worked fine, just like in the case of $SO(3)$. But when I tried to turn the formulas "inside out," I got nonsense. The resulting math simply didn't add up. I tried to fiddle with the formulas but could not find my way around the problem. I had to consider the very real possibility that this construction only worked for $SO(3)$ and not for more general Kac–Moody algebras. There was no way of knowing for sure whether the problem had a solution, and if so, whether a solution could be obtained using the available means. I just had to work as hard as I could and hope for the best.

A week passed, and it was time to meet Fuchs again. I was planning to tell him about my calculations and ask for advice. When I arrived at the dacha, Fuchs told me that his wife had to go to Moscow to run some errands, and he had to take care of his two young daughters.

"But you know what," he said, "Feigin was here yesterday, and he was all excited about the stuff you told us last week. Why don't you go visit him – his dacha is just fifteen minutes away. I told him that I would send you his way today, so he is expecting you."

He explained to me the directions, and I went to Feigin's dacha.

Feigin was indeed expecting me. He greeted me warmly and introduced me to his charming wife, Inna, and his three kids: two energetic boys Roma and Zhenya, ages eight and ten, and an adorable two-year old daughter Lisa. I could not know at that moment that I would be very close to this wonderful family for many years to come.

Feigin's wife offered us some tea and pie, and we sat down on the terrace. It was a beautiful summer afternoon, rays of sun protruding between the leafy trees, birds chirping – idyllic countryside. But of course, the conversation quickly turned to the Wakimoto construction.

It turned out that Feigin was also thinking about it, and along similar lines. At the beginning of our conversation, in fact, we were basically completing each other's sentences. It was a special feeling: he understood me completely, and I understood him.

I started telling him about my failure to generalize the construction to other Kac–Moody algebras. Feigin listened intently, and after sitting quietly for a while, thinking about this, he drew my attention to an important point that I had missed. In trying to generalize Wakimoto's construction, we need to find a proper generalization of the sphere – the manifold on which $SO(3)$ acts by symmetries. In the case of $SO(3)$, this choice is practically unique. But for other groups there are many choices. In my calculations, I had taken it for granted that the natural generalizations of the sphere were the so-called projective spaces. But that was not necessarily the case; the fact that I wasn't getting anywhere with this could just be because my choice of spaces was poor.

As I explained above, at the end of the day, I needed to turn the formulas "inside out." The whole construction hinged on the expectation that, miraculously, the resulting formulas would still be sound. This is what happened in Wakimoto's case, for the simplest group $SO(3)$. My calculations indicated that for the projective spaces, this was not the case, but this didn't mean that a better construction couldn't be found. Feigin suggested I try instead the so-called flag manifolds.[2]

The flag manifold for the group $SO(3)$ is the familiar sphere, so for other groups these spaces may be viewed as natural substitutes for the sphere. But flag manifolds are richer and more versatile than the projective spaces, so there was a chance that an analogue of the Wakimoto construction would work for them.

It was already getting dark, time to go home. We agreed to meet again the following week, and then I waved good-bye to Feigin's family and went back to the train station.

On the train ride home, in an empty train car, with its open windows letting in the warm summer air, I couldn't stop thinking about the problem. I had to try to do it, right there and then. I pulled out a pen and a pad and started writing the formulas for the simplest flag manifold. The old train car, making a staccato noise, was shaking back and forth, and I couldn't hold my pen steady, so the formulas I was writing were all over the place. I could hardly read what I was writing. But in the midst of this chaos, there was a pattern emerging. Things definitely worked better for the flag manifolds than for the projective spaces that I had tried, unsuccessfully, to tame the previous week.

A few more lines of computations, and... Eureka! It was working. The "inside out" formulas worked as nicely as in Wakimoto's work. The construction generalized beautifully. I was overwhelmed with joy: this was the real deal. I did it, I found new free-field realizations of Kac–Moody algebras!

The next morning I checked my calculations carefully. Everything worked out. There was no phone at Feigin's dacha, so I couldn't call him and tell him about my new findings. I started writing them down in a form of a letter, and when we met the following week, I told him about the new results.

This was the beginning of our work together. He became my teacher, mentor, advisor, friend. I addressed him at first as Boris Lvovich, in the old-fashioned Russian way, including the patronymic name. But later he insisted that I switch to the more informal Borya.

I've been incredibly lucky with my teachers. Evgeny Evgenievich showed me the beauty of mathematics and made me fall in love with it. He also helped me learn the basics. Fuchs saved me after the MGU entrance exam catastrophe and jump-started my faltering mathematical career. He led me through my first serious math project, which gave me confidence in my abilities, and steered me to an exciting area of research on the interface of math and physics. Finally, I was ready for the big leagues. Borya proved to be the best advisor that I could possibly dream of at that stage of my journey. It was as though my mathematical career was getting turbocharged.

Borya Feigin is undoubtedly one of the most original mathematicians of his generation in the entire world, a visionary who has the deepest sense of mathematics. He guided me into the wonderland of modern math, full of magic beauty and sublime harmony.

Now that I've had students of my own, I appreciate even more what Borya has done for me (and what Evgeny Evgenievich and Fuchs did for me earlier). It's hard work being a teacher! I guess in many ways it's like having children. You have to sacrifice a lot, not asking for anything in return. Of course, the rewards can also be tremendous. But how do you decide in which direction to point students, when to give them a helping hand and when to throw them in deep waters and let them learn to swim on their own? This is art. No one can teach you how to do this.

Borya cared deeply for me and my development as a mathematician. He never told me what to do, but talking with him and learning from him always gave me a sense of direction. Somehow, he was able to make sure that I always knew what I wanted to do next. And, with him by my side, I always felt confident that I was on the right track. I was very fortunate to have him as my teacher.

It was already the beginning of the fall semester of 1987, my fourth year at Kerosinka. I was nineteen, and my life had never been more exciting. I was still living in the dorm, hanging out with friends, falling in love... I was also keeping up with my studies. By then, I was skipping most of my classes and studying for the exams on my own (occasionally, just a few days prior to the exam). I was still getting straight A's – the only exception being a B in Marxist Political Economy (shame on me).

I kept secret from most people the fact that I had a "second life" – which took up most of my time and energy – my mathematical work with Borya.

I would usually meet with Borya twice a week. His official job was at the Institute for Solid State Physics, but he did not have to do much there, and only had to show up once a week. On other days, he would work at his mother's apartment, which was ten minutes' walk from his home. It was also close to Kerosinka and to my dorm. This was our usual meeting place. I would come in late morning or early afternoon, and we would work on our projects, sometimes all day. Borya's mother would come from work in the evening and feed us dinner, and often we would leave together around nine or ten o'clock.

As our first order of business, Borya and I wrote a short summary of our results and sent it to the journal *Russian Mathematical Surveys*. It was published within a year, pretty fast by the standards of math journals.[3] Having gotten this out of the way, Borya and I focused on developing our project further. Our construction was powerful, and it opened up many new directions of research. We used our results to understand better representations of Kac–Moody algebras. Our work also enabled us to develop a free-field realization of two-dimensional quantum models. This allowed us to make calculations in these models that were not accessible before, which soon made physicists interested in our work.

Those were exciting times. On the days Borya and I weren't meeting, I was working on my own – in Moscow during the week, at home on the weekends. I continued going to the Science Library and devouring more and more books and articles on closely related subjects. I was living and eating and drinking this stuff. It was as though I was immersed in this beautiful parallel universe, and I wanted to stay there, getting ever deeper into this dream. With every new discovery, every new idea, this magical world was becoming more and more my home.

But in the fall of 1988, as I entered the fifth, and last, year of my studies at Kerosinka, I was pulled back to reality. It was time to start thinking about the future. Though I was at the top of my class, my prospects looked bleak. Anti-Semitism ruled out graduate school and the best jobs available to graduates. Not having a *propiska*, residency in Moscow, complicated things even more. The day of reckoning was coming.

Chapter 12

Tree of Knowledge

Even though I knew that I would never be allowed to pursue a career in academia, I continued doing mathematics. Mark Saul talks about this in his article[1] (referring to me by the diminutive form of my first name, Edik):

> What impelled Edik and others to continue, like so many salmon swimming upstream? There was every indication that the discrimination they faced at the university level would continue into their professional lives. Why then should they prepare themselves so intensively and against such odds for a career in mathematics?

I was not expecting to receive anything in return other than the pure joy and passion of intellectual pursuit. I wanted to dedicate my life to mathematics simply because I loved doing it.

In the stagnant life of the Soviet period, talented youth could not apply their energy in business; the economy had no private sector. Instead, it was under tight government control. Likewise, communist ideology controlled intellectual pursuit in the spheres of humanities, economics, and social sciences. Every book or scholarly article in these areas had to start with quotations of Marx, Engels, and Lenin and unequivocally support the Marxist point of view of the subject. The only way to write a paper on foreign philosophy, say, would be to present it as a condemnation of the philosophers' "reactionary bourgeois views." Those who did not follow these strict rules were themselves condemned and persecuted. The same applied to art, music, literature, and cinema. Anything that could be even remotely considered as critical of the Soviet society, politics, or lifestyle – or simply deviated from the canons of "socialist realism" – was summarily censored. The writers, composers, and directors who dared to follow their artistic vision were banned, and their work was shelved or destroyed.

Many areas of science were also dominated by the party line. For example, genetics was banned for many years because its findings were deemed to contradict the teachings of Marxism. Even linguistics was not spared: after Stalin, who considered himself an expert in this subject (as well as many others), wrote his infamous essay *On Certain Questions of Linguistics*, this whole field was reduced to interpreting that largely meaningless treatise. Those who did not follow it were repressed.

In this environment, mathematics and theoretical physics were oases of freedom. Though communist apparatchiks wanted to control every aspect of life, these areas were just too abstract and difficult for them to understand. Stalin, for one, never dared to make any pronouncements about math. At the same time, Soviet leaders also realized the importance of these seemingly obscure and esoteric areas for the development of nuclear weapons, and that's why they did not want to "mess" with these areas. As the result, mathematicians and theoretical physicists who worked on the atomic bomb project (many of them reluctantly, I might add) were tolerated, and some even treated well, by Big Brother.

Thus, on the one hand, mathematics was abstract and inexpensive, and on the other hand, it was useful in the areas that the Soviet leaders cared deeply about – especially defense, which ensured the regime's survival. That's why mathematicians were largely allowed to do their research and were not subjected to the constraints that were imposed on other fields (unless they tried to meddle in politics, as with the "Letter of the 99," which I mentioned earlier).

I believe that this was the main reason why so many talented young students chose mathematics as their profession. This was one area in which they could engage in free intellectual pursuit.

But passion and joy of doing mathematics notwithstanding, I needed a job. Because of this, in parallel with my main mathematical research work, which I did in secret with Borya, I had to do some "official" research at Kerosinka.

My advisor at Kerosinka was Yakov Isaevich Khurgin, who was a professor at the Department of Applied Mathematics and one of the most charismatic and revered faculty members. A former student of Gelfand, Yakov Isaevich was then in his late sixties, but he was one of the "coolest" professors we had. Because of his engaging teaching style and sense of humor, his classes had the highest student attendance. Even though I was skipping most lectures starting from my third year, I always tried to come to his lectures on probability theory and statistics. I started to work with him in my third year.

Yakov Isaevich was very kind to me. He made sure that I was treated well and, whenever I needed help, he was always there for me. For example, when I had some issues at my dorm, he used his leverage to intervene. Yakov Isaevich was a smart man who learned well how to "work the system": even though he was Jewish, he occupied a prestigious position at Kerosinka, as a full professor and the head of a laboratory that did work in areas ranging from oil exploration to medicine.

He was also a popularizer of mathematics, having written several best-selling books about mathematics for non-specialists. I especially liked one of them, entitled *So What?* It is about his collaboration with scientists, engineers, and doctors. Through dialogues with them, he explains in accessible and entertaining ways interesting mathematical concepts (mostly, concerning probability and statistics, his main areas of expertise) and their applications. The title of the book is meant to represent the sense of curiosity with which a mathematician approaches real-life problems. These books and his passion for making mathematical ideas accessible to the public have greatly inspired me.

For many years, Yakov Isaevich worked with medical doctors, mostly urologists. His original motivation was personal. He was enrolled as a student at *Mekh-Mat* when he was called to the front lines of World War II, where he caught a serious kidney disease in the frigid trenches. This was actually lucky for him because he was taken to the hospital, and this saved his life – most of his classmates who were with him in the army were killed in battle. But from that point on, he had to deal with kidney problems. In the Soviet Union medicine was free, but the quality of medical services was low. In order to get a good treatment, one had to have a personal connection with a doctor or have a bribe to offer. But Yakov Isaevich had something else to offer, which very few people could: his expertise as a mathematician. He used it to befriend the best specialists in urology in Moscow.

This was a great deal for him because whenever his kidneys were acting out, he would get the best treatment by top urologists at the best Moscow hospital. And this was a great deal for the doctors too because he would help them analyze their data, which often revealed interesting and previously unknown phenomena. Yakov Isaevich used to say that doctors' thinking was well adapted to analyzing particular patients and making decisions on a case-by-case basis. But this also made it sometimes difficult for them to focus on the big picture and try to find general patterns and principles. This is where mathematicians become useful because our thinking is entirely different: we are trained to look for and analyze these kinds of general patterns. Yakov Isaevich's doctor friends appreciated this.

When I became his student, Yakov Isaevich enlisted me in his medical projects. Ultimately, in the about two and a half years that I worked with Yakov Isaevich, we developed three different projects in urology. The results were used by three young urologists for their doctoral theses. (In Russia, there was a further degree in medicine after M.D., at the level of Ph.D., which required writing a thesis containing original medical research.) I became a co-author of publications in medical journals and even co-authored a patent.

I remember well the beginning of the first project. Yakov Isaevich and I went to see a young urologist Alexei Velikanov, the son of one of Moscow's top physicians. Yakov Isaevich was a longtime friend (and patient) of the older Velikanov, who asked Yakov Isaevich to help his son. Alexei showed us a huge sheet of paper with various data collected from about a hundred patients who had had a prostate adenoma removed (this is a benign tumor of the prostate frequently found in older men). The data included various characteristics, such as blood pressure and other test results, before and after the surgery. He was hoping to use these data to come up with some conclusions about when surgery was more likely to be successful, thereby enabling him to make a set of recommendations as to when to remove the tumor.

He needed help analyzing the data and hoped we could do it. As I learned later, this was a typical situation. Doctors, engineers, and others would often expect that mathematicians have some magic wand that enables them to quickly derive conclusions from whatever data they had collected. Of course, this is wishful thinking. We do know some powerful methods of statistical analysis, but very often these methods cannot be applied because the data are not precise or because there are different types of data: some objective and some subjective (describing how the patients "feel," for example); or some quantitative, such as blood pressure and heartbeat rate, and some qualitative, such as "yes" or "no" answers to some specific questions. It is very difficult, if not outright impossible, to feed such inhomogeneous data into a statistical formula.

On the other hand, sometimes asking the right questions may allow one to realize that some of the data is irrelevant and should be simply thrown away. My experience is that only about 10–15 percent of the information that the doctors collected was ever used when they made the diagnosis or treatment recommendations. But if you asked them, they would never tell you this directly. They would insist that all of it is useful and would even try to come up with some scenario or other in which they would take this information into account. It would take a while to convince them that actually in all of those cases they ignored most of the data and made the decision based on very few essential criteria.

Of course, sometimes there were questions that could be answered by simply feeding the data into some statistical program. But working on these projects, I gradually came to realize that we mathematicians are most useful for doctors not so much because of our knowledge of these statistical programs (after all, this isn't by itself very difficult, and anyone can learn it), but because of our ability to formulate the right questions and then to go through a cold and unbiased analysis to get the answers. It is really this "mathematical mindset" that seems to be most useful to those who are not trained to think as mathematicians.

In my first project, this approach helped us to weed out irrelevant data and then find some non-trivial connections, or correlations, between the remaining parameters. This wasn't easy and took us a few months, but we were happy with the results. We wrote a joint paper about our findings, and Alexei used them in his doctoral thesis. Yakov Isaevich and I were invited to the thesis defense, along with another Kerosinka student, Alexander Lifshitz, my good friend, who also worked on this project.

I remember how at the thesis defense one of the doctors asked for the name of the computer program used to derive these results, and Yakov Isaevich answered that the names were "Edward and Alexander." This was true: we did not use a computer, instead doing all computations by hand or with a simple calculator. The main point was not to calculate (that was the easy part) – it was to ask the right questions. An eminent surgeon, present at the defense, then made a comment to the effect that it was very impressive that mathematics turns out to be so useful in medicine, and perhaps will become even more useful in the years to come. Our work was received well by the medical community, and Yakov Isaevich was pleased.

Soon afterward, he asked me to work on another project in urology that had to do with kidney tumors (for another doctoral thesis), which I was also able to complete successfully.

The third, and last, medical project I worked on was the most interesting one for me. A young doctor, Sergei Arutyunyan – who also needed help to analyze his data for a thesis – and I had a great rapport. He was working with patients whose immune systems were rejecting transplanted kidneys. In such situation the doctor has to make a quick decision whether to fight for the kidney or remove it, with far-reaching consequences: if they kept the kidney, the patient could die, but if they removed it, the patient would need another one, which would be very difficult to find.

Sergei wanted to find a way to tell which recommendation was statistically most viable, based on quantitative ultrasound diagnostics. He had much experience in this area and collected a lot of data. He hoped that I could help him to analyze this data and come up with meaningful objective criteria for decision-making that could be useful to other doctors. He told me that no one had yet been able to do this; most doctors thought this was impossible and preferred to rely on their own *ad hoc* approaches.

I looked at the data. Like in our previous projects, there were about forty different parameters measured for each patient. During our regular meetings, I would ask Sergei pointed questions, trying to figure out which of these data were relevant and which weren't. But this was hard. Like other doctors, he would give his answers based on specific cases, which was not very helpful.

I decided to use a different approach. I thought, "This man makes these kinds of decisions every day, and obviously he is very good at it. What if I manage to learn to 'be him'? Even if I don't know much about the medical aspects of the problem, I could try to learn his methodology following his decision-making process, and then I could use this knowledge to come up with a set of rules."

I suggested that we play a kind of a game.[2] Sergei had collected data on approximately 270 patients. I chose, randomly, the data for thirty of them and put aside the rest. I would take the history of each of these randomly chosen patients and have Sergei, who was sitting at the opposite corner of the office, ask me questions about the patient, which I would answer by consulting the file. My goal in all this was to try to understand the pattern of his questions (even if I could not possibly know the meaning of these questions as well as he did). For example, sometimes he would ask different questions, or the same questions, but in a different order. In such a case, I would interrupt him: "Last time you did not ask this. Why are you asking it now?"

And he would explain, "Because for the last patient the volume of the kidney was so and so, and this ruled out this scenario. But for the current patient it is so and so, and so this scenario is quite possible."

I would make notes of all this and try to internalize this information as much as possible. Even so many years later, I can picture it well: Sergei sitting in a chair in the corner of his office, deep in thought, puffing on a cigarette (he was a chain-smoker). It was fascinating to me to try deconstructing the way he thought – it was kind of like trying to undo a jigsaw puzzle to find out what the essential pieces were.

Sergei's answers gave me extremely valuable information. He would always arrive at the diagnosis after no more than three or four questions. I would then compare it with what actually happened to each patient. He was always spot on.

After a couple dozen cases, I could already make the diagnosis myself, following the simple set of rules that I learned while interrogating him. After half a dozen more, I was practically as good as he was in predicting the outcome. There was in fact a simple algorithm at play that Sergei was following in most cases.

Of course, there were always a handful of cases in which the algorithm would not be useful. But even if one could derive effectively and quickly the diagnosis for 90 to 95 percent of the patients, this would already be quite an achievement. Sergei told me that in the existing literature on the subject of ultrasound diagnostics, there was nothing of this sort.

After completing our "game," I derived an explicit algorithm that I've drawn as a decision tree below. From each node of the tree there are two edges down to other nodes; the answer to a specific question at the first node dictates which of the next two possible nodes the user should go to. For example, the first question is about the index of peripheral resistance (PR) of the blood vessel inside the transplant. This was a parameter Sergei himself had come up with in his research. If its value was greater than 0.79, then it was highly likely that the kidney was being rejected, and the patient required immediate surgery. In this case, we move to the black node on the right. Otherwise, we move to the node on the left and ask the next question: what is the volume (V) of the kidney? And so on. Each patient's data therefore gives rise to a particular path on this tree. The tree terminates after four or fewer steps (it is not important to us at the moment what the remaining two parameters, TP and MPI, were). The terminal node contains the verdict, as shown on this picture: the black node means "operate" and the white node means "do not operate."

I ran the data of the remaining 240 or so patients, whose files I had put aside, through the algorithm. The agreement was remarkable. In about 95 percent of the cases, it led to an accurate diagnosis.

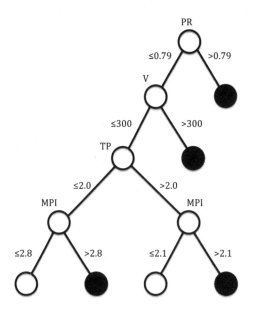

The algorithm described in simple terms essential points of the thought process of a doctor making the decision, and it showed which parameters describing the patient's condition were most relevant to the diagnosis. There were only four of them, narrowing down the initial slate of forty or so. For example, the algorithm showed the importance of the index of peripheral resistance that Sergei had developed, measuring the flow of blood through the kidney. That this parameter played such an important role in the decision-making was, by itself, an important discovery. All of this could be used in further research in this area. Other doctors could apply the algorithm to their patients, test, and perhaps fine-tune it to help make it more efficient.

We wrote a paper about this, which became the basis for Sergei's doctoral thesis, and applied for a patent that was approved a year later.

I was proud of my work with Yakov Isaevich, and he of me. Despite our good relationship, however, I kept my "other" mathematical life – my work with Fuchs and Feigin and all of that – secret from him as I did from most other people. It was as though applied mathematics was my spouse, and pure mathematics was my secret lover.

Still, when the time came that I had to look for a job, Yakov Isaevich told me that he would try to hire me as an assistant at his lab at Kerosinka. That in turn would allow me to become a Ph.D. student there a year later, which would open a clear path to my employment in the foreseeable future. This sounded like an excellent plan, but there were many obstacles, not least that, as my dad had been warned when he went to Kerosinka before I applied there, I would again have to confront anti-Semitism.

Of course, Yakov Isaevich was well aware of this. He had been at Kerosinka for several decades and knew how everything worked. He had actually been hired by Rector Vinogradov himself, whom Yakov Isaevich held in high regard.

Questions of my appointment would be handled by mid-level bureaucrats, not by Vinogradov, and those guys would be sure to close all doors to anyone whose last name sounded Jewish, but Yakov Isaevich knew how to work the system. At the beginning of the spring semester of my last year at Kerosinka, when the question of my employment became urgent, he had typed a letter appointing me to his laboratory. He carried the letter with him in his briefcase, so that, should he get a chance to talk to Vinogradov about me personally, he would be well-prepared.

The opportunity soon presented itself. One day he bumped into Vinogradov as he was entering Kerosinka. Vinogradov was pleased to see him and asked, "How are you doing, Yakov Isaevich?"

"Terribly," replied Yakov Isaevich grimly (he could be a good actor).

"What happened?"

"We have done wonderful things at my lab in the past, but we can't do this anymore. I can't get new talent. I have this great student who is graduating this year, but I am unable to hire him."

I suppose Vinogradov wanted to show Yakov Isaevich who was the boss – which is what Yakov Isaevich's goal was – so he said, "Don't worry, I will take care of this."

At which point, Yakov Isaevich produced my appointment letter. Vinogradov had no choice but to sign it.

Normally, this letter had to be signed by a dozen people before landing on Vinogradov's desk: the heads of the local Komsomol organization and Communist Party, and all kinds of other bureaucrats. They would surely find a way to stall the process so that this would never happen. But now it already had Vinogradov's signature! So what could they do? He *was* the boss, and they couldn't possibly disobey his wishes. They would grind their teeth and stall for a while, but eventually they would all give up and sign. You should have seen their faces when they saw Vinogradov's signature at the bottom! Yakov Isaevich had played the system brilliantly.

Chapter 13

Harvard Calling

In the midst of all that stress and uncertainty, in March of 1989, a letter arrived from the United States, on the letterhead of Harvard University.

> Dear Dr. Frenkel,
>
> Upon the recommendation of the Department of Mathematics, I would like to invite you to visit Harvard University in the fall of 1989 as the recipient of the Harvard Prize Fellowship.
>
> Sincerely yours,
>
> Derek Bok
>
> President of Harvard University

I had heard about Harvard University before, though I must admit I did not quite realize at the time its significance in the academic world. Still, I was very pleased. Being invited to America as a winner of a fellowship sounded like a big honor. The President of the University personally wrote to me! And he addressed me as a "Doctor," even though I had not yet received my bachelor's degree. (I was then in the last semester of my studies at Kerosinka.)

How did this happen? The word about my work with Borya was spreading. Our first short paper had already been published, and we were finishing three other, longer, papers (all in English). The physicist Lars Brink, visiting Moscow from Sweden, solicited one of them for a volume he was editing. We gave him our paper for the volume and asked him to make twenty or so copies and send them to mathematicians and physicists abroad we thought would be interested in our work. I had found their addresses in published papers that were available at the Moscow Science Library and gave the list to Lars. He kindly agreed to help us because he knew how difficult it would be for us to send copies ourselves. That paper became widely known, in part because of its applications to quantum physics.

This was several years before the usage of the Internet became widespread, but the system of dissemination of scientific literature was quite efficient: authors would circulate typed manuscripts of their articles before publication (they were called preprints). The recipients would then copy and forward them further to their colleagues as well as university libraries. The twenty or so people to whom Lars Brink sent our paper must have done the same.

In the meantime, tremendous changes were happening in the Soviet Union: it was the time of *perestroika*, launched by Mikhail Gorbachev. One of the results of this was that people were allowed to travel abroad with greater freedom. Before then, mathematicians like Feigin and Fuchs received many invitations to attend conferences and visit universities in the West, but foreign travel was closely regulated by the government. Before getting the usual entry visa to another country, one had to get an exit visa, which enabled the person to leave the Soviet Union. Very few of those visas were given out, out of fear that people would not return (and indeed, many of those who were granted exit visas didn't come back). Almost all requests were denied, usually on bogus grounds, and Fuchs once told me that it had been years since he had last tried to get one.

But suddenly, in the fall of 1988, several people were allowed to travel abroad, one of whom was Gelfand. Another, a talented young mathematician and a friend of Borya named Sasha Beilinson, also traveled to the United States to visit his former co-author Joseph Bernstein, who had emigrated a few years earlier and was a professor at Harvard.

In the meantime, some scientists in the West also recognized that changes were coming and tried to use this opportunity to invite scholars from the Soviet Union. One of them was Arthur Jaffe, a renowned mathematical physicist who was then the chairman of the Mathematics Department at Harvard. He decided to create a new visiting position for talented young Russian mathematicians. When Gelfand, who had an honorary degree from Harvard, came to visit in the fall of 1988, Jaffe enlisted his help in order to convince the president, Derek Bok, whom Gelfand knew personally, to provide funding and support for this program (some funding was also provided by Landon Clay, who later founded the Clay Mathematics Institute). Jaffe called it the Harvard Prize Fellowship.

Once the program was in place, the question was whom to invite, and Jaffe polled various mathematicians for suggestions. Apparently, my name was mentioned by several people (including Beilinson), and this was the reason that I was selected among the first four recipients of the award.

The letter from President Bok was quickly followed by a longer letter from Jaffe himself, which described the terms of the appointment in more detail. I could come for a period between three and five months; I would be a Visiting Professor but wouldn't have any formal obligations except for giving occasional lectures about my work; Harvard would pay for my travel, housing, and living expenses. Practically the only thing Harvard wasn't providing was a Soviet exit visa. Fortunately, and to my great surprise, I got it in a month.

Arthur Jaffe wrote in his letter that I could come as early as the end of August and stay till the end of January, but I chose to stay for three months, which was the minimum period my stay specified in the letter. Why? Well, I had no intention of emigrating to the U.S., and I was planning to come back. Besides, I was feeling guilty that I would have to take a leave from the job at Kerosinka that Yakov Isaevich had won for me with such an effort.

After I got my exit visa and it became clear that my trip was becoming reality, I had to come clean with Yakov Isaevich and tell him about my "extracurricular activities": my mathematical work with Feigin and the invitation from Harvard. Naturally, he was very surprised. He was sure that I was devoting all my energy to the medical projects on which I was working with him. His first reaction was quite negative.

"And who is going to work in my lab if you go to Harvard?" he asked.

At this point Yakov Isaevich's wife, Tamara Alekseevna, who always warmly welcomed me at their home, came to my rescue:

"Yasha, you are talking nonsense," she said. "The kid got invited to Harvard. This is great news! He should definitely go, and when he comes back he will continue working with you."

Yakov Isaevich reluctantly agreed.

The summer months went by quickly and the date of my departure, September 15, 1989, came. I flew from Moscow to New York's JFK airport and then to Boston. Jaffe could not meet me at the airport personally, but he sent a graduate student to pick me up. I was taken to the two-bedroom apartment that the Math Department rented for me and another Harvard Prize Fellow, Nicolai Reshetikhin, who was arriving a few days later. It was in the Botanic Gardens, an apartment complex owned by Harvard, less than a ten-minute walk from Harvard Yard. Everything looked different and exciting.

It was late at night when I got to my apartment. Jet-lagged, I immediately went to sleep. The next morning I went to a nearby farmer's market and got some produce. Back home, I started making a salad and realized that I didn't have salt. There was none at the apartment, so I had to eat the salad unsalted.

As soon as I finished it, the door bell rang. It was Arthur Jaffe. He proposed to give me a quick tour around the city in his car. That was really cool – a twenty-one-year-old kid being driven around the city by the chairman of the Harvard Math Department. I saw Harvard Yard, the Charles River, beautiful churches, and the skyscrapers in downtown Boston. The weather was beautiful. I was very impressed by the city.

On the way back from the two-hour journey, I told Arthur that I needed to buy salt, and he said, "No problem, I'll take you to a nearby supermarket."

He took me to the Star Market on Porter Square and said that he would wait for me in his car.

This was the first time that I ever was at a supermarket, and it was a startling experience. At that time there were shortages of food in Russia. In my hometown, Kolomna, one could only buy bread, milk, and basic vegetables, like potatoes. For other food one had to travel to Moscow, and even in Moscow the best one could hope for was some low-grade mortadella sausage and cheese. Every weekend, when I came home from Moscow, I brought with me some food for my parents. So seeing aisle after aisle packed with all kinds of food at the supermarket was absolutely unbelievable.

"How do you ever find anything here?" I thought. I started walking down the aisles looking for salt, but I couldn't find any. I guess I must have been slightly dizzy from the abundance of stuff in the supermarket; at any rate, I didn't even notice the signs at the top. I asked someone working at the supermarket: "Where is salt?" but I couldn't understand a word of what he said. My English was good enough to give a math lecture, but I didn't have any experience with everyday colloquial English. The heavy Bostonian accent didn't make understanding any easier.

Half an hour passed by, and I was really getting desperate, lost in the Star Market like in a giant labyrinth. Finally, I came across a package of salt mixed with garlic. "Good enough," I said to myself, "let's get out of here." I paid and came out of the store. Poor Arthur got worried – what the hell was this kid doing in there for forty-five minutes? – so he already started looking for me.

"Losing myself in the abundance of capitalism," I thought.

My adaptation to America had begun.

The other two recipients of the Harvard Prize Fellowship arriving in the fall semester were Nicolai Reshetikhin, with whom I was sharing my apartment (he arrived a week later), and Boris Tsygan.* They were both ten years my senior, and they both had already made seminal contributions to mathematics. I knew about their work but had never met them before in person. During that first semester, we bonded and became friends for life.

Nicolai, or Kolya as he is affectionately known to many, was from St. Petersburg. He was already famous as one of the inventors of the so-called quantum groups, which are generalizations of the ordinary groups. More precisely, quantum groups are certain deformations of Lie groups – the mathematical objects we talked about earlier. These quantum groups are now as ubiquitous as Lie groups in many areas of mathematics and physics. For example, Kolya and another mathematician, Vladimir Turaev, used them to construct invariants of knots and three-dimensional manifolds.

Borya Tsygan was a longtime collaborator of Boris Feigin, my teacher. Originally from Kiev, Ukraine, Tsygan had a big idea when he was fresh out of college, which led to a breakthrough in the field of "non-commutative geometry." Like other Jewish mathematicians, he was kept from going to graduate school after college. Because of this, after he graduated from the university, he had to work at a heavy machinery plant in Kiev, spending all day surrounded by loud machines. Nevertheless, it was in these, less-than-perfect, conditions that he made his discovery.

People tend to think that mathematicians always work in sterile conditions, sitting around and staring at the screen of a computer, or at a ceiling, in a pristine office. But in fact some of the best ideas come when you least expect them, possibly through annoying industrial noise.

Walking around Harvard Yard and taking in the old-fashioned architecture of the brick buildings, the statue of Harvard, the spires of ancient churches, I couldn't help but feel the exclusivity of this place, with its long tradition of the pursuit of knowledge and never-ending fascination with discovery.

Harvard's Department of Mathematics was housed in the Science Center, a modern-looking building just outside of Harvard Yard. It had the look of a giant alien space ship that just happened to land in Cambridge, Massachusetts, and decided to stay there. The Math Department occupied three floors in it. Inside, offices mixed with common areas complete with coffee machines and comfortable couches. There was also a well-designed in-house math library,

---

*Vera Serganova, the fourth recipient of the Harvard Prize Fellowship, came in the spring.

and even a ping-pong table. All this created a homey atmosphere, and even in the middle of the night you could find plenty of people there, young and old – working, reading in the library, pacing nervously in the corridors, engaged in a lively conversation, and some just dozing off on the couch... You had the feeling that you would never have to leave this place (and it seemed that some people never did).

The department was quite small in comparison with other schools: the faculty consisted of no more than fifteen permanent professors and ten or so postdocs holding three-year teaching positions. When I arrived, the faculty included some of the greatest mathematicians of our time, such as Joseph Bernstein, Raoul Bott, Dick Gross, Heisuke Hironaka, David Kazhdan, Barry Mazur, John Tate, and Shing-Tung Yau. Meeting them and learning from them was the opportunity of a lifetime. I have fond memories of the charismatic Raoul Bott, a gray-haired friendly giant, then in his late sixties, pulling me aside in the corridor and asking, in his booming voice, "How are you doing, young man?"

There were also thirty or so graduate students who all had tiny cubicles on the middle floor.

The three Russians – Kolya, Borya, and I – were greeted warmly by everyone. Although we were the beginning of a tidal wave of Russian scientists who swept American universities in the ensuing years, it was still highly unusual to have visitors from the Soviet Union in those days. Still, after a week or so in Cambridge, I felt like I blended right in. Everything seemed so natural and cool. I bought myself the hippest jeans and a Sony Walkman (remember, this was 1989!), and I was walking around town wearing headphones, listening to the coolest tracks. To a stranger I would have looked like a typical twenty-one-year-old student. My conversational English still left something to be desired. To improve it, every day I would buy *The New York Times* and read it, with a dictionary, for at least an hour (deciphering some of the most arcane English words one could find, as I later realized). I also got addicted to late-night TV.

David Letterman's show (which was then starting at 12:35 am on NBC) was my favorite. The first time I watched it, I couldn't understand a single word. But somehow it was clear that this was *my* show, that I would really enjoy it, if only I could understand what the host was saying. So this provided some extra motivation for me. I would stubbornly watch the show every night, and little by little I started to understand the jokes, the context, the background. This was my way of discovering American pop culture, and I was devouring every bit of it. Some nights, when I had to go to sleep early, I would videotape the show and watch it in the morning while having breakfast. The Letterman show became something of a religious ritual for me.

Although the other fellows and I did not have any formal obligations, we came to the department every day to work on our projects, to talk to people, and to attend seminars, of which there were plenty. The two professors I talked to the most were the two Russian expatriates: Joseph Bernstein and David Kazhdan. Both are amazing mathematicians, former students of Gelfand, and close friends of each other, but you couldn't imagine more different temperaments. Joseph is quiet and unassuming. If asked a question, he would listen quietly, take his time to think, and would often say that he didn't know the answer, but would still tell you what he thought about the subject. His explanations were crystal clear and down-to-earth, and often he would actually explain the answer that he claimed he didn't know. He always made you feel that you didn't have to be a genius to understand all this stuff – a great feeling for an aspiring young mathematician.

David, on the other hand, is a dynamo – extremely sharp, witty, and quick. In his encyclopedic knowledge, panache, and occasional display of impatience, he recalls his teacher Gelfand. At seminars, if he thought that the speaker wasn't explaining the material well, he would simply walk up to the blackboard, wrestle the chalk from the speaker's hand, and just take over – that is, if he was interested in the subject. Otherwise, he could simply doze off. It was quite rare to hear him say "I don't know" in response to a question – he really knows pretty much everything. I've spent long hours talking to him over the years and have learned a lot. Later on, we collaborated on a joint project, which was a rewarding experience.

In my second week at Harvard I had another fateful encounter. Besides Harvard, there is another, lesser-known, school in Cambridge, usually referred to by an abbreviation of its name... MIT. (I am kidding, of course!) There has always been a bit of a rivalry between Harvard and MIT, but in fact the two Math Departments are very closely connected. It's not unusual, for example, for a Harvard student to have an MIT professor as an advisor and vice versa. Students from each school often attend classes offered by the other school.

Sasha Beilinson, Borya Feigin's friend and co-author, was appointed as a professor at MIT, and I was attending the lectures he was giving there. At the first lecture, someone pointed out to me a handsome man in his mid-forties sitting a couple of rows away. "This is Victor Kac." Wow! This was the creator of the Kac–Moody algebras and many other things, whose works I had been studying for several years.

After the lecture, we were introduced. Victor greeted me warmly and told me that he wanted to learn more about my work. I was thrilled when he invited me to speak at his weekly seminar. I ended up giving three talks at his seminar, on three consecutive Fridays. These were my first seminars in English, and I think I did a decent job: the attendance was high, people seemed to be interested, and they asked many questions.

Victor took me under his wing. We would often meet in his spacious office at MIT, talk about math, and he would often invite me over to his house for dinner. We subsequently worked together on several projects.

About a month after my arrival, Borya Feigin came to Cambridge as well. Sasha Beilinson sent him an invitation to visit MIT for two months. I was happy that Borya came to Cambridge: he was my teacher, and we were very close. We also had a number of ongoing math projects, and this was a great opportunity to work on them. I did not realize at first that his visit would also throw my life into a great turmoil.

The news that the door to the West was now open, and mathematicians could freely travel and visit universities in the United States and elsewhere, quickly spread through the Moscow mathematical community. Many people decided to seize this opportunity and move permanently to America. They started sending applications to various universities and calling their colleagues in the U.S., telling them that they were looking for jobs. Since no one knew for how long this policy of "openness" would continue (most people expected that after a few months the borders would be sealed again), this created a sort of frenzy in Moscow – all conversations led to the same question: "How best to get out?"

And how could it be otherwise? Most of these people had to deal with anti-Semitism and various other obstacles in the Soviet Union. They could not find employment in academia and had to do mathematics on the side. And, although the mathematical community in the Soviet Union was very strong, it was largely isolated from the rest of the world. There were great opportunities for professional development in the West that simply did not exist in the Soviet Union. How could one expect these people to be loyal to the country that rejected them and tried to prevent them from working in the field they loved, when the opportunities of better life abroad presented themselves?

When he came to the U.S., Borya Feigin saw right away that a great "brain drain" was coming, and nothing could stop it. In Russia, the economy was falling apart, with shortages of food everywhere, and the political situation was becoming more and more unstable. In America, there was a much higher standard of living, an abundance of everything, and the life of the academics just seemed so comfortable. The contrast was immense. How could one possibly convince anyone to go back to the Soviet Union after experiencing all of this firsthand? The exodus of the overwhelming majority of the top tier mathematicians from Russia – or anyone, really, who could find a job – seemed inevitable, and it was going to happen very quickly.

Nevertheless, Borya resolved to return to Moscow, despite the fact that he had been struggling with anti-Semitism all his life and had no illusions about the situation in the Soviet Union. He was accepted to Moscow University as an undergraduate (in 1969, when he applied, some Jewish students were still accepted), but he was not allowed to enter the Ph.D. program there. He had to enroll at the university in the provincial city of Yaroslavl to get his Ph.D. He then had a great difficulty finding a job, until he was able to get a position at the Institute for Solid State Physics. Still, Borya found this rush to the exit disturbing. He thought it was morally wrong to leave Russia *en masse* like this at the time of great upheaval, like rats abandoning a sinking ship.

Borya was extremely saddened that the great Moscow Mathematical School was soon to be no more. The tight-knit community of mathematicians that he was living in for so many years was about to evaporate in front of his very eyes. He knew that he would soon be practically alone in Moscow, deprived of the greatest pleasure of his life: doing mathematics together with his friends and colleagues.

Naturally, this became the main theme of my conversations with Borya. He tried to convince me that I should go back and not succumb to what he deemed to be the mass hysteria possessing those who were trying to escape to the West. He was also worried that I would not be able to become a good mathematician in the U.S. American "consumer society," he thought, can kill one's motivation and work ethics.

"Look, you've got talent," he would say to me, "but it needs to be developed further. You have to work hard, the way you were working in Moscow. Only then can you realize your potential. Here, in America, this is impossible. There are too many distractions and temptations. Life here is all about fun, enjoyment, instant gratification. How can you possibly focus on your work here?"

I wasn't buying his argument, at least not entirely. I knew that I had a strong motivation to do mathematics. But I was only twenty-one, and Borya, fifteen years my senior, was my mentor. I owed him everything I had achieved as a mathematician. His words gave me pause – what if he was right?

The invitation to Harvard was a turning point in my life. Only five years earlier, I was failed at the MGU exam, and it looked like my dream of becoming a mathematician was irreparably shattered. Coming to Harvard was my vindication, a reward for all the hard work I did in Moscow in those five years. But I wanted to keep on moving, making new discoveries. I wanted to become the best mathematician I could be. I looked at the invitation to Harvard as just one stage in a long journey. It was an advance: Arthur Jaffe and others believed in me and gave me this opportunity. I couldn't let them down.

In Cambridge, I was lucky to have the support of wonderful mathematicians like Victor Kac who encouraged me and helped me in every way they could. But I also sensed jealousy from some of my colleagues: Why was this guy given so much so soon? What has he done to deserve it? I felt compelled to fulfill my promise, to prove to everybody that my first mathematical works were not a fluke, that I could do better and bigger things in mathematics.

Mathematicians form a small community, and like all humans, they gossip about who is worth what. In my short time at Harvard I had already heard enough stories about prodigies who burned out early. I'd heard some unforgiving comments about them, things like, "Remember so-and-so? His first works were so good. But he hasn't done anything nearly as important in the last three years. What a shame!"

I was terrified that in three years they would say this about me, so I constantly felt under pressure to produce and to succeed.

Meanwhile, the economic situation in the Soviet Union was rapidly deteriorating, and the outlook was very uncertain. Observing all of this from the inside and convinced that I had no future in the Soviet Union, my parents took to calling me at regular intervals urging me *not* to come back. In those days, it was very difficult (and expensive) to call the U.S. from the Soviet Union. My parents were afraid that their home phone was tapped, and because of that, they would travel to the Moscow central post office and place a call from there. Such a trip would take them almost an entire day. But they were determined, even though they missed me terribly, to do everything in their power to convince me to stay in America. They were absolutely sure that this was in my best interests.

Borya also had my best interests in mind, but he took his stance in part on moral grounds. He was going against the grain, and I admired him for this.

But I also had to admit that he could afford to do so because of his relatively comfortable situation in Moscow (though that was soon going to change, and he would be forced to spend a few months a year abroad – mostly, in Japan – to provide for his family). My situation was entirely different: I had no place to stay in Moscow, and only a temporary *propiska*, the right to live there. Though Yakov Isaevich had secured a temporary job as an assistant for me at Kerosinka, it provided a meager salary that would barely be enough to rent a room in Moscow. Because of anti-Semitism, getting enrolled in a graduate school would be an uphill battle, and my future employment prospects looked even bleaker.

At the end of November, Arthur Jaffe called me to his office and offered to extend my stay at Harvard till the end of May. I had to make a decision quickly, but I was torn. I enjoyed my lifestyle in Boston. I felt this was my place to be. With Harvard and MIT, Cambridge was one of the premier centers of mathematics. Some of the world's most brilliant minds were here, and I could just knock on their doors, ask them questions, learn from them. There were also plenty of seminars where pretty much all exciting discoveries were reported soon after they were made. I was surrounded by the brightest students. This was the most stimulating environment for a young aspiring mathematician one could imagine. Moscow used to be such a place, but not any more.

But this was the first time I was away from home for such a long time. I missed my family and friends. And Borya, my teacher, who was the closest person to me in Cambridge, was adamant that I should go back in December, as planned.

Every morning I woke up terrified, thinking "What should I do?" In retrospect, the decision seems like a no-brainer. But with so many different forces colliding, all at the same time, making that decision wasn't easy. Finally, after some anguished deliberations, I decided to follow the advice of my parents and stay, and I told Jaffe about it. My friends Reshetikhin and Tsygan did the same.

Borya was unhappy about this, and I felt that I had let him down. It was a moment of sadness and great uncertainty when I saw him off at the Logan Airport, going back to Moscow, in mid-December. We didn't know what the future held for either of us; we didn't even know whether we would be able to meet again any time soon. I had ignored Borya's advice. But I was still afraid that I might fulfill his fears.

Chapter 14

Tying the Sheaves of Wisdom

The spring semester brought more visitors to Harvard, one of whom, Vladimir Drinfeld, changed the direction of my research, and in many ways, my mathematical career. And it all happened because of the Langlands Program.

I had heard about Drinfeld before. He was only thirty-six at the time, but already a legend. Six months after we met, he was awarded the Fields Medal, one of the most prestigious prizes in mathematics, considered by many as an equivalent of the Nobel Prize.

Drinfeld published his first math paper at the age of seventeen, and by age twenty he was already breaking new ground in the Langlands Program. Originally from Kharkov, Ukraine, where his father was a well-known math professor, Drinfeld studied at Moscow University in the early 1970s. (At that time, Jews also had trouble gaining admission to MGU, but a certain percentage of Jewish students was admitted.) By the time he received his college degree from the MGU, he was already world-renowned for his work, and he was accepted to the graduate school, which was extraordinary for a Jewish student. His advisor was Yuri Ivanovich Manin, one of the world's most original and influential mathematicians.

Even Drinfeld, however, was not able to escape anti-Semitism entirely. After getting his Ph.D., he was unable to get a job in Moscow and had to spend three years at a provincial university in Ufa, an industrial city in the Ural Mountains. Drinfeld was reluctant to go to Ufa, not least because there were no mathematicians there working in the areas he cared about. But as the result of his stay in Ufa, Drinfeld wrote an important work in the theory of integrable systems, a subject that was quite far from his interests, together with a local mathematician Vladimir Sokolov. The integrable systems they created are now known as the Drinfeld–Sokolov systems. After three years in Ufa, Drinfeld was finally able to secure a job in his hometown, at the Kharkov Institute for Low Temperature Physics. This was a relatively comfortable job, and he could stay close to his family, but being in Kharkov, Drinfeld was isolated from the Soviet mathematical community, which was concentrated in Moscow and, to a lesser extent, St. Petersburg.

Despite all this, working essentially alone, Drinfeld kept producing marvelous results in diverse areas of math and physics. In addition to proving important conjectures within the Langlands Program and opening a new chapter in the theory of integrable systems with Sokolov, he also developed the general theory of quantum groups (originally discovered by Kolya Reshetikhin and his co-authors), and many other things. The breadth of his contributions was staggering.

Attempts were made to hire Drinfeld in Moscow. I've been told that physicist Alexander Belavin, for example, tried to bring Drinfeld to the Landau Institute for Theoretical Physics near Moscow. To raise the chances of success, Belavin and Drinfeld solved together an important problem of classification of solutions to the "classical Yang–Baxter equation," which many physicists were interested in at the time. Their paper was published to much acclaim in Gelfand's journal *Functional Analysis and Applications* (I believe that this was the longest article ever published by Gelfand, which is saying a lot about its importance). It was that work that led Drinfeld to the theory of quantum groups, which revolutionized many areas of mathematics. Alas, none of these hiring plans worked. Anti-Semitism and Drinfeld's lack of *propiska* in Moscow were a deadly combination. Drinfeld remained in Kharkov, visiting Moscow only rarely.

Drinfeld was invited to visit Harvard in the spring of 1990, and this proved to be serendipitous for me. He arrived in late January. Having heard all the legends about him, I was a bit intimidated at first, but he turned out to be extremely nice and generous. Soft-spoken, carefully weighing his words, Drinfeld was also the very model of clarity when he talked about mathematics. When he explained things to you, he did not try to do it in a self-aggrandizing way, as if he was unveiling a big mystery which you would never be able to fully understand on your own (which unfortunately is the case for some of our colleagues, who shall remain nameless). On the contrary, he was always able to put things in the simplest and clearest possible way, so after he explained something to you, you felt like you'd known it all along.

More importantly, Drinfeld told me right away that he was very interested in my work with Feigin, which he wanted to use for his new project related to the Langlands Program.

Let's recall from Chapter 9 the three columns of André Weil's Rosetta stone:

Number Theory   *Curves over Finite Fields*   *Riemann Surfaces*

The Langlands Program was originally developed within the left and the middle columns: number theory and curves over finite fields. The idea of the Langlands Program is to set up a relation between the representations of a Galois group and automorphic functions. The concept of the Galois group makes perfect sense in the left and the middle columns of the Rosetta stone, and there are suitable automorphic functions that can be found in another area of mathematics called harmonic analysis.

Prior to Drinfeld's work, it was not clear whether there was an analogue of the Langlands Program for the right column, the theory of Riemann surfaces. The means to include Riemann surfaces started to emerge in the early 1980s in Drinfeld's work, which was followed by the French mathematician Gérard Laumon. They realized that it was possible to make a geometric reformulation of the Langlands Program that makes sense for both the middle column and the right column of André Weil's Rosetta stone.

In the left and the middle columns of the Rosetta stone, Langlands Program relates the Galois groups and the automorphic functions. The question is then to find the right analogues of the Galois groups and the automorphic functions in the geometric theory of Riemann surfaces. We have already seen in Chapter 9 that in the geometric theory the role of the Galois group is played by the fundamental group of a Riemann surface. But we left the geometric analogues of the automorphic functions unexplored.

It turns out that the right geometric analogues are not functions, but what mathematicians call *sheaves*.

To explain what they are, let's talk about numbers. We have natural numbers: 1,2,3,..., and of course they have many uses. One is that they measure dimensions. As we discussed in Chapter 10, a line is one-dimensional, a plane is two-dimensional, and for any natural number n we have an n-dimensional flat space, also known as a vector space.[1] Now imagine a world in which natural numbers are replaced by vector spaces; that is, instead of number 1 we have a line, instead of number 2 we have a plane, and so on.

Addition of numbers is replaced in this new world by what mathematicians call the direct sum of vector spaces. Given two vectors spaces, each with its own coordinate system, we create a new one, which combines the coordinates of the two vector spaces, so its dimension is the sum of two dimensions. For example, a line has one coordinate, and a plane has two. Combining them, we obtain a vector space with three coordinates. This is our three-dimensional space.

Multiplication of natural numbers is replaced by another operation on vector spaces: given two vectors spaces, we produce a third one, called their tensor product. I will not give a precise definition of the tensor product here; the important point is that if the two vector spaces we start with have dimensions m and n, then their tensor product has dimension $m \cdot n$.

Thus, we have operations on vectors spaces that are analogous to the operations of addition and multiplication of natural numbers. But this parallel world of vector spaces is so much richer than the world of natural numbers! A given number has no inner structure. Number 3, for example, taken by itself, has no symmetries. But a three-dimensional space does. In fact, we have seen that any element of the Lie group $SO(3)$ gives rise to a rotation of the three-dimensional space. The number 3 is a mere shadow of the 3-dimensional space, reflecting only one attribute of this space, its dimensionality. But this number cannot do justice to other aspects of the vector spaces, such as its symmetries.

In modern math, we create a new world in which numbers come alive as vector spaces. Each of them has a rich and fulfilling personal life, and they also have more meaningful relations with each other, which cannot be reduced to mere addition and multiplication. Indeed, we can subtract 1 from 2 in only one way. But we can embed a line in a plane in many different ways.

Unlike natural numbers, which form a set, vector spaces form a more sophisticated structure, which mathematicians call a category. A given category has "objects," such as vector spaces, but in addition, there are "morphisms" from any object to any other object.[2] For example, morphisms from an object to itself in a given category are essentially the symmetries of that object that are allowed within this category. The language of categories therefore enables us to focus not on what the objects consist of, but on how they interact with each other. Because of that, the mathematical theory of categories turns out to be particularly well-adapted to computer science.[3] The development of functional programming languages, such as Haskell, is just one example of a myriad of recent applications.[4] It seems inevitable that the next generations of computers will be based more on category theory than on set theory, and categories will enter our daily lives, whether we realize it or not.

The paradigm shift from sets to categories is also one of the driving forces of modern math. It is referred to as *categorification*. We are, in essence, creating a new world, in which the familiar concepts are elevated to a higher level. For example, numbers get replaced by vector spaces. The next question is: what should become of functions in this new world?

To answer this question, let's revisit the notion of a function. Suppose we have a geometric shape, like a sphere or a circle, or the surface of a donut. Call it S. As we already discussed before, mathematicians refer to such shapes as manifolds. A function f on a manifold S is a rule that assigns to each point s in S a number, called the value of the function f at the point s. We denote it by $f(s)$.

An example of a function is temperature, with our manifold S simply the three-dimensional space we live in. At each point s we can measure the temperature, which is a number. This gives us a rule assigning to each point a number, so we get a function. Likewise, barometric pressure also gives us a function.

For a more abstract example, let S be the circle. Each point of the circle is determined by an angle, which, as before, we will call φ. Let f be the sine function. Then the value of this function at the point of the circle corresponding to the angle φ is $\sin(\varphi)$. For example, if $\varphi = 30$ degrees (or $\pi/6$, if we measure angles in radians rather than degrees), then the value of the sine function is $1/2$. If $\varphi = 60$ degrees (or $\pi/3$), then it is $\sqrt{3}/2$, and so on.

Now let's replace numbers by vector spaces. So a function will become a rule that assigns to each point s in a manifold S, not a number, but a vector space. Such a rule is called a *sheaf*. If we denote a sheaf by the symbol $\mathscr{F}$, then the vector space assigned to a point s will be denoted by $\mathscr{F}(s)$.

Thus, the difference between functions and sheaves is in what we assign to each point of our manifold S: for functions, we assign numbers to points, and for sheaves, we assign vector spaces. For a given sheaf, these vector spaces can be of different dimensions for different points s. For example, on the picture below, most of these vector spaces are planes (that is, two-dimensional vector spaces), but there is one that is a line (that is, a one-dimensional vector space). Sheaves are categorifications of functions, in the same way as vector spaces are categorifications of numbers.

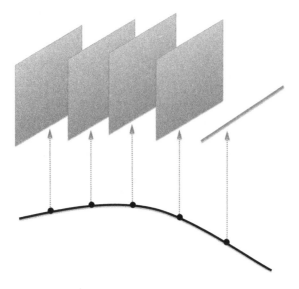

Though this is beyond the scope of this book, a sheaf is actually more than just a disjoint collection of vector spaces assigned to the points of our manifold. The fibers of a given sheaf at different points have to be related to each other by a precise set of rules.[5]

What matters to us at the moment is that there is a deep analogy between functions and sheaves, discovered by the great French mathematician Alexander Grothendieck.

The influence of Grothendieck on modern mathematics is virtually unparalleled. If you ask who was the most important mathematician of the second half of the twentieth century, many mathematicians will say without hesitation: Grothendieck. Not only did he almost singlehandedly create modern algebraic geometry, he also transformed the way we think about mathematics as a whole. The dictionary between functions and sheaves, which we use in the geometric reformulation of the Langlands Program, is an excellent example of the profound insights that are characteristic of Grothendieck's work.

To give you a gist of Grothendieck's idea, I recall the notion of a finite field from Chapter 8. For each prime number p, there is a finite field with p elements: $\{0, 1, 2, ..., p-1\}$. As we discussed, these p elements comprise a numerical system with operations of addition, subtraction, multiplication, and division modulo p, which obey the same rules as the corresponding operations on the rational and real numbers.

But there is also something special about this numerical system. If you take any element of the finite field $\{0, 1, 2, ..., p-1\}$ and raise it to pth power – in the sense of the arithmetic modulo p that we discussed earlier – you will get back the same number! In other words,

$$a^p = a \quad \text{modulo} \quad p.$$

This formula was proved by Pierre Fermat, the mathematician who came up with Fermat's Last Theorem. Unlike the proof of the latter, though, the proof of the above formula is fairly simple. It could even fit in the margin of a book. I've put it in the back of this one.[6] To distinguish this result from Fermat's Last Theorem (sometimes also referred as Fermat's Great Theorem), it is called Fermat's little theorem.

For example, set $p = 5$. Then our finite field is $\{0, 1, 2, 3, 4\}$. Let's raise each of them to the 5th power. Surely, 0 to any power is 0, and 1 to any power is 1, so no surprises here. Next, let's raise 2 to the 5th power: we then get 32. But $32 = 2 + 5 \cdot 6$, so modulo 5 this is 2 – we get back 2, as promised. Let's take the 5th power of 3: we get 243, but this is $3 + 5 \cdot 48$, so it is 3 modulo 5. Again, we get back the number we started with. And finally, let's try the same with 4: its 5th power is 1024, which is 4 modulo 5. Bingo! I encourage you to check that $a^3 = a$ modulo 3, and $a^7 = a$ modulo 7 (for larger primes you might need a calculator to verify Fermat's little theorem).

What's also remarkable is that a similar equation forms the basis of the RSA encryption algorithm widely used in online banking.[7]

This formula $a^p = a$ is more than a neat discovery – it means that the operation of raising numbers to the pth power, sending a to a^p, is an element of the Galois group of the finite field. It is called the Frobenius symmetry, or simply the *Frobenius*. It turns out that the Galois group of the finite field of p elements is generated by this Frobenius.[8]

Let's go back to Grothendieck's idea. We start in the middle column of Weil's Rosetta stone. Then we study curves over finite fields and more general manifolds over finite fields. These manifolds are defined by systems of polynomial equations such as

$$y^2 + y = x^3 - x^2,$$

which we talked about in Chapter 9.

Suppose that we have a sheaf on such a manifold. It is a rule assigning to each point of the manifold a vector space, but there is actually more structure. The notion of a sheaf is defined in such a way that any symmetry of the numerical system over which our manifold is defined – which is in this case a finite field – gives rise to a symmetry of this vector space. In particular, the Frobenius, which is an element of the Galois group of the finite field, necessarily gives rise to a symmetry (such as a rotation or a dilation) of this vector space.

Now, if we have a symmetry of a vector space, we can produce a number out of it. There is a standard technique for doing this. For example, if our vector space is a line, then the symmetry of this space that we obtain from the Frobenius will be a dilation: each element z will be transformed to Az for some number A. Then the number we assign to this symmetry is just A. And for the vector spaces of dimension greater than one, we take what's called the trace of the symmetry.[9] By taking the trace of the Frobenius on the space $\mathscr{F}(s)$, we assign a number to the point s.

The simplest case is that the Frobenius acts as the identity symmetry on the vector space. Then its trace is equal to the dimension of the vector space. So in this case, by taking the trace of the Frobenius, we assign to a vector space its dimension. But if the Frobenius is not the identity, this construction assigns to a vector space a more general number, which is not necessarily a natural number.

The upshot is that if we have a manifold S over a finite field $\{0, 1, 2, ..., p-1\}$ (which is what happens if we are in the middle column of Weil's Rosetta stone) and we have a sheaf $\mathscr{F}$ on S, then to each point s of S we can assign a number. This gives us a function on S. Therefore we see that in the middle column of Weil's Rosetta stone we have a way to go from sheaves to functions.

Grothendieck called this a "sheaves–to–functions dictionary." It is a curious sort of dictionary, however. Relying on the procedure described above, we obtain a passage from sheaves to functions. Furthermore, natural operations on sheaves are parallel to natural operations on functions. For example, the operation of taking the direct sum of two sheaves, defined similarly to the direct sum of two vector spaces, is parallel to the operation of taking the sum of two functions.

But there is no natural way to go back from functions to sheaves.[10] It turns out we can do this only for some functions, not for all. But if we can do it, then this sheaf will carry a lot of additional information that the function did not have. This information can then be used to get to the heart of that function. A remarkable fact is that most of the functions that appear in the Langlands Program (in the second column of Weil's Rosetta stone) do come from sheaves.

Mathematicians studied functions, one of the central notions in all of mathematics, for centuries. This is a concept that we can grasp intuitively by thinking of the temperature or barometric pressure. But what people didn't recognize before Grothendieck is that if we are in the context of manifolds over finite fields (such as curves over a finite field), we can go beyond functions and work with sheaves instead.

Functions were, if you will, the concepts of archaic math, and sheaves are the concepts of modern math. Grothendieck showed that in many ways sheaves are more fundamental; the good ol' functions are their mere shadows.

This discovery greatly stimulated progress in mathematics in the second half of the twentieth century. The reason is that sheaves are much more vital and versatile objects, with a lot more structure. For example, a sheaf can have symmetries. If we elevate a function to a sheaf, we can exploit these symmetries, and this way we can learn a lot more than we could ever learn using functions.

What is especially important for us is that sheaves make sense both in the middle column and in the right column of Weil's Rosetta stone. This opens a path to moving the Langlands Program from the middle to the right column.

In the right column of the Rosetta stone, we consider manifolds that are defined over the complex numbers. For example, we consider Riemann surfaces such as the sphere or the surface of a donut. In this setting, the automorphic functions that appear in the left and middle columns of Weil's Rosetta stone do not make much sense. But sheaves do make sense. So once we replace functions by sheaves in the middle column (which we can do because we have

Grothendieck's dictionary), we regain the analogy between the middle and the right columns of Weil's Rosetta stone.

Let's summarize: when we pass from the middle column of Weil's Rosetta stone to the right column, we have to make some adjustments to both sides of the relation envisioned by the Langlands Program. That's because the notions of Galois group and automorphic functions don't have immediate counterparts in the geometry of Riemann surfaces. First, the Galois group finds its analogue in the fundamental group of a Riemann surface, as explained in Chapter 9. Second, we use the Grothendieck dictionary and instead of automorphic functions, we consider sheaves that satisfy properties analogous to the properties of the automorphic functions. We call them automorphic sheaves.

This is illustrated by the following diagram, in which we have three columns of the Rosetta stone, and the two rows in each column contain the names of the objects on the two sides of the Langlands relation specific to that column.

Number theory	Curves over finite fields	Riemann surfaces
Galois group	Galois group	fundamental group
automorphic functions	automorphic functions or automorphic sheaves	automorphic sheaves

The question is then how to construct these automorphic sheaves. This proved to be a very difficult problem. In the early 1980s, Drinfeld proposed the first such construction in the simplest case (building on an earlier unpublished work of Pierre Deligne). Drinfeld's ideas were further developed by Gérard Laumon a few years later.

When I met Drinfeld, he told me that he had come up with a radically new method to construct automorphic sheaves. But the new construction he envisioned was conditional on a certain conjecture which he thought I could derive from my work with Feigin on Kac–Moody algebras. I couldn't believe it: my work could be useful for the Langlands Program?

The chance that I could do something related to the Langlands Program made me eager to learn everything that was known about it. That spring, I went to Drinfeld's office at Harvard almost every day, and I pestered him with questions about the Langlands Program, which he patiently answered. He would also ask me about my work with Feigin, the details of which were critical for what he was trying to do. The rest of the day I would devour anything that I could find on the Langlands Program at the Harvard library. The subject was so alluring, I tried to fall asleep every night as quickly as I could, so that morning would come sooner and I would immerse myself ever deeper into the Langlands Program. I knew I was embarking on one of the most important projects of my life.

Something else happened near the end of the spring semester that threw me right back to the Kafkaesque experience of my entrance exams at Moscow University.

One day, Victor Kac called me at home in Cambridge and told me that someone invited Anatoly Logunov, the President (or Rector, as he was called) of Moscow University to give a lecture at MIT's physics department. Kac and many of his colleagues were outraged that MIT would give forum to the man directly responsible for the discrimination against Jewish students at the entrance exams to MGU. Kac and the others felt that his actions amounted to a crime, and hence the invitation was scandalous.

Logunov was a very powerful man: he was not only the President of MGU but also director of the Institute of High Energy Physics, member of the Central Committee of the Communist Party of the USSR, and more. But why would someone at MIT invite him? In any case, Kac and several of his colleagues protested and asked for the visit and the lecture to be canceled. After some negotiations, a compromise was found: Logunov would come and give his lecture, but after the lecture there would be a public discussion of the situation at the MGU, and people would have a chance to confront him about the discrimination. It would be something like a town-hall meeting.

Naturally, Kac asked me to come to that meeting to present my story as firsthand evidence of what was happening at MGU under Logonov's leadership. I was a bit reluctant to do this. I was sure that Logunov would be accompanied by "assistants," who would be recording everything. Remember, this was May 1990, more than a year before the failed *putsch* of August 1991 that started the collapse of the Soviet Union. And I was about to go back home for the summer. If I were to say something even mildly embarrassing for such a high-profile Soviet official as Logunov, I could easily get in trouble. At the very least, they could prevent me from leaving the Soviet Union and returning to Harvard. Still, I couldn't deny Kac's request. I knew how important my testimony could be at this meeting, so I told Victor that I would come and, if needed, tell my story. Kac tried to reassure me.

"Don't worry, Edik," he said, "if they put you in jail for this, I'll do everything I can to get you out."

The word about the upcoming event quickly spread, and the lecture hall was packed for Logunov's lecture. People did not come to learn anything from his talk. Everybody knew that Logunov was a weak physicist who made his career trying to disprove Einstein's relativity theory (I wonder why). As expected, the talk – on his "new" theory of gravity – had very little substance. But it was quite unusual in many respects. First of all, Logunov did not speak English and delivered his lecture in Russian, which was simultaneously translated by a tall man in a black suit and tie who spoke perfect English. He might as well have had "KGB" written on his forehead in big block letters. His clone (as in the movie *The Matrix*) was sitting in the audience looking around.

Before the talk, one of Logunov's MIT hosts introduced him in a very peculiar fashion. He projected a slide of the first page of a paper in English, co-authored by Logunov and a few other people and published a decade earlier. I guess the point was to show that Logunov was not a total idiot, but actually he had to his credit some publications in refereed journals. I'd never seen anyone being introduced in this way. It was clear that Logunov was not invited to MIT for his scientific brilliance.

There were no protests at the lecture, though Kac had distributed among the audience members photocopies of some damning documents. One of them was the transcript of a fellow with a Jewish last name from a decade earlier. He had A's on all subjects and yet during his last year at MGU he was expelled for "academic failure." A short note added to the transcript informed the reader that this student was spotted by specially dispatched agents at the Moscow synagogue.

After the lecture, people went to another room and sat around a large rectangular table. Logunov was sitting on one side, close to an end, flanked by the two plain-clothed "assistants," who were translating, and Kac and other accusers were sitting directly across the table from them. I sat quietly with a few friends at the opposite end of the table, to Logunov's side, so he wasn't paying any attention to us.

At first, Kac and others spoke and said that they had heard many stories about Jewish students not being admitted to MGU. They asked Logunov if he, as the Rector of Moscow University, would comment on this. Of course, he flatly denied everything, no matter what they said to him. At some point, one of the plain-clothed guys said, in English, "You know, Professor Logunov is a very modest person, so he would never tell you this. But I will. He has actually helped many Jewish people with their careers."

The other plain-clothed guy then said to Kac and others, "You should either put up or shut up. If you have any concrete cases that you want to talk about, bring them up. Otherwise, Professor Logunov is very busy and he has other things to attend to."

At this point, of course, Kac said, "Actually, we do have a concrete case that we want to talk to you about," and he gestured toward me.

I stood up. Everyone turned to me, including Logunov and his "assistants," their faces betraying some anxiety. I was now facing Logunov directly.

"Very interesting," said Logunov in Russian – this was to be translated into English for everybody – and then added quietly to his assistants, but I could hear: "Don't forget to take down his name."

I have to say, it was a bit scary, but I had reached the point of no return. I introduced myself and said, "I was failed at the entrance exams to the *Mekh-Mat* six years ago."

And then I briefly described what had happened at the exams. The room went quiet. This was a "concrete" eyewitness account from one of the victims of Logunov's policy, and there was no way he could deny that this had happened. The two assistants rushed to limit the damage.

"So you were failed at the MGU. And where did you apply after this?" one of them asked.

"I went to the Institute of Oil and Gas," I said.

"He went to Kerosinka," translated the assistant to Logunov, who then vigorously nodded *yes* – of course, he knew that this was one of the few places in Moscow where students like me could be accepted.

"Well," continued the assistant, "Perhaps, the competition at the Oil and Gas Institute was not as stiff as at MGU. Maybe that's why you got into one, but not the other?"

This was false: I knew for a fact there was very little competition at *Mekh-Mat* among those who were not discriminated against. I'd been told that getting one B and three C's at the four exams was sufficient to get in. Entrance exams to Kerosinka were, on the contrary, very competitive. At this point Kac interjected: "While a student, Edward did some groundbreaking mathematical work and was invited as a Visiting Professor to Harvard at the age of twenty-one, less than five years after he was failed at MGU. Are you going to suggest that the competition for the Harvard position was also lower than at the entrance exam to MGU?"

Long silence. Then, suddenly, Logunov became very animated.

"I am outraged by this!" he yelled. "I will investigate and punish those responsible for failing this young man. I will not allow this kind of things to happen at MGU!"

And he went on like this for a few minutes.

What could one possibly say to this? No one at the table believed that Logunov's outburst was genuine and that he would really do anything. Logunov was very clever. By expressing his feigned outrage about one case, he deflected a much bigger issue: that of thousands of other students who were ruthlessly failed as the result of a carefully orchestrated discrimination policy that was clearly approved by the top brass at MGU, including the Rector himself.

We couldn't possibly bring up those cases at that meeting and prove that there was a concerted policy of anti-Semitism at the entrance exams to *Mekh-Mat*. And while there was a certain measure of satisfaction that I was able to face my tormentor directly and force him to admit that I was indeed wronged by his subordinates, we all knew that the bigger question remained unanswered.

Logunov's hosts, who were clearly embarrassed by all the negative publicity surrounding his visit, wanted to get this over with as quickly as possible. They adjourned the meeting and whisked him away. He was never invited back.

Chapter 15

A Delicate Dance

In the fall of 1990, I became a Ph.D. student at Harvard, which I had to do to move from Visiting Professor to something more permanent. Joseph Bernstein agreed to be my official advisor. By then, I already had more than enough material for a Ph.D. thesis, and Arthur Jaffe got the Dean to waive the usual two-year enrollment requirement for me so that I could get my Ph.D. in one year. Because of that, my "demotion" from professor to a graduate student didn't last very long.

In fact, I wrote my Ph.D. thesis about a new project, which I completed during that year. It all started from my discussions with Drinfeld on the Langlands Program in the spring. Here's one of them, in the form of a screenplay.

FADE IN:

INT. DRINFELD'S OFFICE AT HARVARD

DRINFELD is pacing in front of the blackboard. EDWARD, sitting in a chair, is taking notes. A tea cup is on the desk by his side.

                    DRINFELD
          So, the Shimura-Taniyama-Weil conjecture
          gives us a link between cubic equations and
          modular forms, but Langlands went much
          farther than this. He envisioned a more

DRINFELD (CONT'D)

general relation, in which the role of modular forms is played by the automorphic representations of a Lie group.

EDWARD

What's an automorphic representation?

DRINFELD
(after a long pause)

The precise definition is not important now. And anyway, you can read it in a book. What's important is that it is a representation of a Lie group G - for example, the group $SO(3)$ of rotations of a sphere.

EDWARD

OK. And what are these automorphic representations related to?

DRINFELD

Well, that's the most interesting part: Langlands predicted that they should be related to representations of the Galois group in another Lie group.[1]

EDWARD

I see. You mean this Lie group is not the same group G?

DRINFELD

No! It's another Lie group, which is called the Langlands dual group of G.

DRINFELD writes the symbol ^{L}G on the blackboard.

                              EDWARD
          Is the L for Langlands?

                              DRINFELD
                          (hint of a smile)
          Well, Langlands' original motivation was to
          understand something called L-functions, so
          he called this group an L-group...

                              EDWARD
          Let me see if I understand this. For every
          Lie group G, there is another Lie group
          called ^{L}G, correct?

                              DRINFELD
          Yes. And it appears in the Langlands
          relation, which looks schematically as
          follows.

DRINFELD draws a diagram on the blackboard:[2]

representations of the Galois group in ^{L}G	$\longleftrightarrow$	automorphic representations of the group G

                              EDWARD
          I don't understand it... at least, not
          yet. But let me ask you a simpler
          question: what's the Langlands dual group
          of $SO(3)$, say?

                              DRINFELD
          That's pretty easy: It is the double cover
          of $SO(3)$. Have you seen the cup trick?

                         EDWARD
        The cup trick? Oh yes, I remember...

FADE TO:

INT. A HOUSE PARTY OF HARVARD GRADUATE STUDENTS

A dozen or so students, in their early to mid-twenties, are
talking, drinking beer and wine. EDWARD is talking to a STUDENT.

                         STUDENT
        Here's how it works.

The STUDENT takes a plastic cup with wine and puts it on the open
palm of her right-hand. She then starts rotating her palm and
arm (as shown on the series of photographs below). After she
makes a full turn (360 degrees), her arm gets twisted. Keeping
the cup upright, she continues rotating, and after another full
turn - surprise! - her arm and the cup come back to the initial,
untwisted, position.[3]

                         ANOTHER STUDENT
                I've heard Filipinos have a traditional
                wine dance, in which they do this with both
                hands.[4]

He picks up two cups of beer and tries to rotate them both at
once, but his hands are unsteady, and he quickly spills beer from
both cups. Everybody is laughing.

FADE TO:

INT. BACK TO DRINFELD'S OFFICE

> DRINFELD
> The trick illustrates the fact that there
> is a closed path on the group $SO(3)$ which
> is non-trivial, but if we traverse this
> path twice, we get a trivial path.[5]

> EDWARD
> Oh, I see. The first full turn of the cup
> twists your arm when you do this, and
> that's like a non-trivial path on $SO(3)$.

He picks up a cup of tea from the table and goes through the
motion of the first twist.

> EDWARD
> You would think that making the second turn
> would twist the arm even more. Instead,
> the second turn untwists the arm.

EDWARD completes the motion.

> DRINFELD
> Exactly.[6]

> EDWARD
> What does this have to do with the
> Langlands dual group?

> DRINFELD
> The Langlands dual group of $SO(3)$ is a
> double cover of $SO(3)$, so...

Cup trick (read left to right, top to bottom). Photos by Andrea Young.

                    EDWARD
So for each element of $SO(3)$, there are two
elements of the Langlands dual group.

                    DRINFELD
Because of this, this new group[7] won't have
any non-trivial closed paths.

                    EDWARD
So passing to the Langlands dual group is a
way to get rid of this funny twisting?

                    DRINFELD
That's right.[8] At first glance, this might
seem like a minor difference, but in fact
it has major effects, such as the
difference in behavior between the building
blocks of matter, like electrons and
quarks, and the particles that carry
interactions between them, like photons.
For more general Lie groups, the difference
between the group and its Langlands dual
group is even more pronounced. In fact, in
many cases, there is no apparent link
between the two dual groups.

                    EDWARD
Why does the dual group appear in the
Langlands relation? Seems like magic...

                    DRINFELD
We don't really know.

FADE TO BLACK

The Langlands duality sets up a pairwise relationship between Lie groups:

for every Lie group G, there is a Langlands dual Lie group ^LG, and the dual of ^LG is G itself.[9] It is surprising enough that the Langlands Program relates two different types of objects (one from number theory and one from harmonic analysis), but the fact that two dual groups, G and ^LG, appear on the two sides of this relation, as shown on the diagram on p. 168, is mind-boggling.

We have talked about the Langlands Program connecting different continents of the world of mathematics. By way of analogy, let's say those continents were Europe and North America, and we had a way to match every person in Europe with a person in North America, and vice versa. Moreover, suppose that under this relation, various attributes, such as weight, height, and age, matched perfectly, but genders got switched: every man was matched with a woman, and vice versa. Then this would be like the switch between a Lie group and its Langlands dual group under the relation predicted by the Langlands Program.

This switch is in fact the most mysterious aspect of the Langlands Program. We know several mechanisms that describe how the dual group appears, but we still don't understand *why* this happens. That ignorance is one of the reasons we try to expand the ideas of the Langlands Program to other fields of mathematics (through Weil's Rosetta stone) and then to quantum physics, as we will see in the next chapter. We want to find more examples of the appearance of the Langlands dual group and hope that this will give us more clues about why it happens, and what it means.

Let's focus now on the right column of Weil's Rosetta stone, which concerns Riemann surfaces. As we established in the previous chapter (see the diagram on p. 161), in the version of the Langlands relation that plays out in this column, the cast of characters has "automorphic sheaves" in the role of automorphic functions (or automorphic representations) associated to a Lie group G. It turns out that these automorphic sheaves "live" on a certain space attached to a Riemann surface X and the group G, called the moduli space of G-bundles on X. It's not important to us at the moment what it is.[10] On the other side of the relation, the role of the Galois group is played by the fundamental group of this Riemann surface, as we have seen in Chapter 9. From the diagram on p. 168, we then find that the geometric Langlands relation (also known as the geometric Langlands correspondence) should schematically look as follows:

representations of the fundamental group of X in ^LG	$\longleftrightarrow$	automorphic sheaves on the moduli space of G-bundles on X

This means that to each representation of the fundamental group in ^LG, we should be able to associate an automorphic sheaf. And Drinfeld had a radically new idea about how to do this.

FADE IN:

INT. DRINFELD'S OFFICE

                    DRINFELD
       So we have to find a systematic way to
       construct these automorphic sheaves. And I
       think representations of Kac-Moody algebras
       can do the trick.

                    EDWARD
       Why is that?

                    DRINFELD
       We are now in the world of Riemann
       surfaces. Such a surface may have a
       boundary, which consists of loops.

DRINFELD draws a picture on the blackboard.

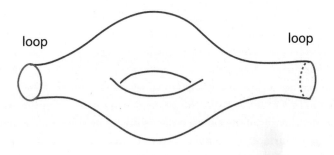

loop   loop

                    DRINFELD
       Loops on a Riemann surface give us a link
       to loop groups and hence to Kac-Moody
       algebras. Using this link, we can convert

DRINFELD (CONT'D)

representations of a Kac-Moody algebra into
sheaves on the moduli space of G-bundles
on our Riemann surface. Let's ignore the
details for now. Schematically, I expect
it to work like this.

He draws a diagram on the blackboard:

| a representation of the fundamental group in ^LG | $\longrightarrow$ | a representation of the Kac-Moody algebra of G | $\longrightarrow$ |

$\longrightarrow$ | an automorphic sheaf on the moduli space of G-bundles |

DRINFELD

The second arrow is clear to me. The real
question is how to construct the first
arrow. Feigin told me about your work on
the representations of Kac-Moody algebras.
I think you can put it to use here.

EDWARD

But then representations of the Kac-Moody
algebra of G should somehow "know" about
the Langlands dual group ^LG.

DRINFELD

That's right.

EDWARD

How is this possible?

DRINFELD

That is a question for you.

FADE TO BLACK

I guess I felt a little like Neo talking to Morpheus in the film *The Matrix*. It was exciting and also a little scary. Will I really be able to say something new about this field?

In order to explain how I approached this problem, I need to tell you about an efficient method of constructing representations of the fundamental group of a Riemann surface. We do this by using differential equations.

A differential equation is an equation that relates a function and its derivatives. As an example, let's look at a car moving on a straight road. The road has one coordinate; let's denote it by x. The position of the car at the moment t in time is then encoded by a function $x(t)$. For example, it could be that $x(t) = t^2$.

The velocity of the car is the ratio of the distance traveled over a small time period Δt to this time period:

$$\frac{x(t + \Delta t) - x(t)}{\Delta t}.$$

If the car were traveling with a constant velocity, it would not matter which time period Δt we take. But if the car is changing its velocity, then a smaller Δt will give us a more accurate approximation for the velocity at the moment t. In order to get the exact, instant value of the velocity at that moment, we have to take the limit of this ratio as Δt goes to 0. This limit is the derivative of $x(t)$. It is denoted by $x'(t)$.

For example, if $x(t) = t^2$, then $x'(t) = 2t$, and more generally, if $x(t) = t^n$, then $x'(t) = nt^{n-1}$. It's not difficult to derive these formulas, but this is not essential to us now.

Many laws of nature may be expressed as differential equations, that is, equations involving functions and their derivatives. For example, Maxwell's equations describing electromagnetism, which we will talk about in the next chapter, are differential equations, and so are the Einstein equations describing the force of gravity. In fact, the majority of mathematical models (be they in physics, biology, chemistry, or financial markets) involve differential equations. Even the simplest questions one can ask about personal finance, such as how to compute compound interest, quickly lead us to differential equations.

Here is an example of a differential equation:

$$x'(t) = \frac{2x(t)}{t}.$$

The function $x(t) = t^2$ is a solution of this equation. Indeed, we have $x'(t) = 2t$ and $2x(t)/t = 2t^2/t = 2t$, so substituting $x(t) = t^2$ into the left- and right-hand sides we obtain the same expression, $2t$. Moreover, it turns out that any solution of this equation has the form $x(t) = Ct^2$, where C is a real number independent from t (C stands for "constant"). For example, $x(t) = 5t^2$ is a solution.

Similarly, the solutions of the differential equation

$$x'(t) = \frac{nx(t)}{t}$$

are given by the formula $x(t) = Ct^n$, where C is an arbitrary real number.

Nothing prevents us from allowing n to be a negative integer here. The equation will still make sense, and the formula $x(t) = Ct^n$ will still make sense, except that this function will no longer be defined at $t = 0$. So let's exclude $t = 0$ from consideration. Once we do this, we can also allow n to be an arbitrary rational number, and even an arbitrary real number.

And now we make an additional step: in the original formulation of this differential equation, we treated t as time, so it was assumed to be a real number. But now let's suppose that t is a complex number, so it has the form $r + s\sqrt{-1}$, where r and s are real numbers. As we discussed in Chapter 9 (see the picture on p. 102), complex numbers may be represented as points on the plane with coordinates r and s. Once we make t complex, $x(t)$ effectively becomes a function on the plane. Well, the plane minus one point, that is. Since we decided that $x(t)$ may not be defined at the point $t = 0$, which is the origin on this plane (with both coordinates, r and s, equal to zero), $x(t)$ is really defined on the plane excluding one point, the origin.

Next, we bring the fundamental group into the game. Elements of the fundamental group, as we discussed in Chapter 9, are closed paths. Let's consider the fundamental group of the plane with one point removed. Then any closed path has a "winding number": this is the number of times the path goes around the removed point. If the path goes counterclockwise, we count this number with the positive sign, and if it goes clockwise, with the negative sign.[11] The closed paths with the winding numbers +1 and -1 are shown on the picture.

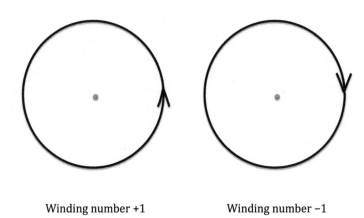

Winding number +1   Winding number −1

A path that spirals around twice and crosses itself to return to where it began would have a winding number of either +2 or -2, and so on for more complicated paths.

Let's go back to our differential equation:

$$x'(t) = \frac{nx(t)}{t},$$

where n is an arbitrary real number and t now takes values in complex numbers. This equation has a solution $x(t) = t^n$. However, there is a surprise in store: if n is not an integer, then as we evaluate the solution along a closed path on the plane and come back to the same point, the value of the solution at the end point will not necessarily be the value we started with. It will get multiplied by a complex number. In this situation, we say that the solution has undergone a *monodromy* along this path.

The statement that something changes when we make a full circle may sound counterintuitive and even self-contradictory at first. But it all depends on what we mean by making a full circle. We may be traversing a closed path and returning to the same point in the sense of a *particular* attribute, such as our position in space. But other attributes may well undergo changes.

Consider this example. Rick met Ilsa at a dinner party on March 14, 2010, and instantly fell in love. Ilsa didn't think much of Rick at first but agreed to go on a date with him anyway. And then on another date. And another. Ilsa started to like Rick; why, he is funny and smart, taking good care of her. Next thing you know, Ilsa was also in love; she even changed her Facebook status to "in a relationship," and so did Rick. Time flew fast, and soon it was March

14 again, the one-year anniversary of the day they first met. From the point of view of the calendar – if we only pay attention to the month and day and ignore the year – Rick and Ilsa had made full circle. But things changed. On the day they met, Rick was in love, and Ilsa wasn't. But a year later, that is no longer the case; in fact, they could be equally in love with each other, or perhaps, Ilsa, head over heels, and Rick just so-so. It is even possible that Rick would have fallen out of love with Ilsa and started secretly seeing someone else. We don't know. What is important to us is that even though they had come back to the same calendar date, March 14, their love for each other may well have changed.

Now, my father tells me that this example is confusing because it seems to suggest that Rick and Ilsa came back to the same point in time, which is impossible. But what I am focusing on are particular attributes: specifically, month and day. In that respect, going from March 14, 2010, to March 14, 2011, is really making a circle.

But maybe it's better to consider instead a closed path in space. So suppose that while they were together, Rick and Ilsa went on a trip around the world. As they were traveling, their relationship was evolving, so when they came back to the same point in space – their hometown – their love for each other may have changed.

In the first case, we have a closed path in time (more precisely, in the month-and-date calendar), and in the second case, a closed path in space. But the conclusions are similar: a relationship may change along a closed path. Both scenarios illustrate a phenomenon that we could call the monodromy of love.

Mathematically, we can represent Rick's love for Ilsa by a number x, and Ilsa's love for Rick by a number y. Then the state of their relationship at each moment would be represented by a point on the plane with coordinates (x, y). For example, in the first scenario, on the day they first met, this was the point $(1, 0)$. But then, as they were moving along a closed path (in time or in space), the position of the point changed. Hence the evolution of their relationship is represented by a trajectory on the xy-plane. The monodromy is simply the difference between the initial point and the end point of this trajectory.

Here is a less romantic example. Suppose you climb a spiral staircase and make a full turn. As far as the projection of your position on the floor is concerned, you have made a full circle. But another attribute – your altitude – has changed: you have moved on to the next level. That's also a monodromy. We can tie this with our first example because the calendar is like a spiral: 365 days of the year is like a circle on the floor, and the year is like the altitude.

Moving from a given date, such as March 14, 2010, to the same date a year later is therefore similar to climbing a staircase.

Let's go back to the solution of our differential equation. A closed path on the plane is like the closed path of your projection on the floor. The value of the solution is like the altitude of your position on the staircase. From this point of view, it should not come as a surprise that the value of the solution as we make a full turn would be different from the initial value.

Taking the ratio of these two values, we obtain the monodromy of the solution along this path. It turns out that we can interpret this monodromy as an element of the circle group.[12] To illustrate this, imagine that you could bend a candy cane so that it would take the shape of a donut. Then follow the red swirl. Moving along the cane is like following a closed path on our plane, the swirl being our solution. When we make a full circle on the cane, the swirl will in general come back to a different point than where it started. That difference is like the monodromy of our solution. It corresponds to a rotation of the cane by some angle.

The computation presented in endnote 12 shows that the monodromy along a closed path with the winding number +1 is the element of the circle group corresponding to the rotation by $360n$ degrees. (For example, if n is 1/6, then to this path we assign the rotation by 360/6 = 60 degrees.) Likewise, the monodromy along the path with the winding number w is the rotation by $360wn$ degrees.

The upshot of this discussion is that the monodromies along different paths on the plane without a point give rise to a representation of its fundamental group in the circle group.[13] More generally, we can construct representations of the fundamental group of any Riemann surface (possibly, with some points removed, as in our case) by evaluating the monodromy of differential equations defined on this surface. These equations are going to be more complicated, but locally, within a small neighborhood of a point on the surface, they all look similar to the one above. Using the monodromy of solutions of even more sophisticated equations, we can construct in a similar way representations of the fundamental group of a given Riemann surface in Lie groups other than the circle group. For example, we can construct representations of the fundamental group in the group $SO(3)$.

Let's return to the problem I was facing: we start with a Lie group G and take the corresponding Kac–Moody algebra. Drinfeld's conjecture required finding a link between representations of this Kac–Moody algebra and representations of the fundamental group in the Langlands dual group $^L G$.

The first step is to replace representations of the fundamental group by suitable differential equations whose monodromy takes values in $^L G$. This makes the question more algebraic and hence closer to the world of Kac–Moody algebras. The kinds of differential equations that are relevant here were introduced earlier (essentially, in the case of a plane without a point, as above) by Drinfeld and Sokolov during the time when Drinfeld was "exiled" to Ufa. Beilinson and Drinfeld subsequently generalized this work to arbitrary Riemann surfaces and called the resulting differential equations "opers." The word "oper" is derived from "operator," but it was also partly a joke, because in Russian it is a slang word for a police officer, like "cop."

In my thesis, building on the work I did in Moscow with Borya, I was able to construct representations of the Kac–Moody algebra of G parametrized by the opers corresponding to the Langlands dual group $^L G$. The existence of a link between the two was nearly miraculous: the Kac–Moody algebra associated to G somehow "knew" about the Langlands dual group $^L G$, as Drinfeld had predicted. This made his plan work according to the following scheme:[14]

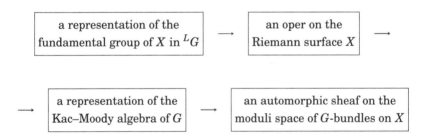

My proof of this result was technically quite involved. I was able to explain *how* the Langlands dual group appeared, but even now, more than twenty years later, I still find mysterious *why* it appears. I solved the problem, but it was ultimately unsatisfying to feel that something just appeared out of thin air. My research since then has been motivated in part by trying to find a more complete explanation.

It often happens like this. One proves a theorem, others verify it, new advances in the field are made based on the new result, but the true understanding of its meaning might take years or decades. I know that even if I don't find the answer, the torch will be passed to new generations of mathematicians who will eventually figure it out. But of course, I would love to get to the heart of it myself.

Beilinson and Drinfeld subsequently used the theorem from my thesis in their beautiful construction of the geometric Langlands relation (in the right column of Weil's Rosetta stone, see the diagram on p. 173). Their spectacular work was the beginning of a new chapter in the Langlands Program, bringing a host of fresh ideas and insights into the subject and expanding it even further.

I later summarized the research I have done in this area (some of it in collaboration with Borya, and some with Dennis Gaitsgory) in my book *Langlands Correspondence for Loop Groups* published by Cambridge University Press.[15] It came out in 2007, exactly twenty years after I wrote the first formulas for the free-field realization of the Kac–Moody algebras on a night train ride home from Borya's dacha, a calculation that – little did I know – began my long journey to the Langlands Program.

As an epigraph to my book, I chose these lines from a 1931 poem by E.E. Cummings, one of my favorite poets:

> Concentric geometries of transparency slightly
> joggled sink through algebras of proud
> inwardlyness to collide spirally with iron arithmethics...

To me, it sounds like a poetic metaphor for what we are trying to achieve in the Langlands Program: a unity of geometry, algebra, and arithmetic (that is, number theory). A modern-day alchemy.

The work of Beilinson and Drinfeld solved some long-standing problems, but it also raised more questions. That's how it is in mathematics: each new result pushes back the veil covering the unknown, but what then becomes known doesn't simply encompass answers – it includes questions we didn't know to ask, directions we didn't know we could explore. And so each discovery inspires us to make new strides and never leaves us satisfied in our pursuit of knowledge.

In May 1991, I attended the graduation ceremony at Harvard. It was an even more special moment for me because the commencement speaker was Eduard Shevardnadze, one of the architects of the *perestroika* in the Soviet Union. He had recently resigned his post as foreign minister in protest against the violence in the Baltic republics, warning against an incipient dictatorship.

Those were turbulent times. We didn't know about all the turmoil that was yet to come: the *coup d'état* in August of that year, the subsequent breakup of the Soviet Union, the immense hardship that most people would have to endure in the course of economic reforms. Nor could we anticipate Shevardnadze's controversial stint as the head of his native Republic of Georgia. But on that glorious day in the sunlit Harvard Yard, I wanted to say "thank you" to the man who helped to free me, and millions of my compatriots, from the communist regime.

I came up to him after his speech and told him I had just received my Ph.D. from Harvard, which wouldn't have been possible without *perestroika*. He smiled and said, in Russian, with his charming Georgian accent, "I am glad to hear this. I wish you great success in your work." He paused and added, as a true Georgian: "And happiness in your personal life."

The next morning I flew to Italy. Victor Kac invited me to a conference he organized in Pisa with his Italian colleague Corrado De Concini. From Pisa, I went to the island of Corsica to attend another meeting, and then to a conference in Kyoto, Japan. These conferences brought together physicists and mathematicians interested in Kac–Moody algebras and their applications to quantum physics. I lectured about the work I had just completed. This was the first time most of the participants heard about the Langlands Program, and they seemed to be intrigued by it. Thinking back to those days, I am amazed how much things have changed since then. The Langlands Program is now considered a cornerstone of modern math and is widely known across diverse disciplines.

This was the first time I had the opportunity to travel around the world. I was discovering different cultures and also realizing how mathematics, our common language, brings us closer together. Everything was new and exciting, the world – a kaleidoscope of infinite possibilities.

Chapter 16

Quantum Duality

We have seen the Langlands Program reverberate through chambers of mathematics, from number theory to curves over finite fields to Riemann surfaces. Even representations of Kac–Moody algebras have gotten into the mix. Through the lens of the Langlands Program, we observe the same patterns, the same phenomena in these diverse mathematical fields. They manifest themselves in different ways, but some common features (such as the appearance of the Langlands dual group) can always be recognized. They point to a mysterious underlying structure – we might say, the source code – of all mathematics. It is in this sense that we speak of the Langlands Program as a Grand Unified Theory of mathematics.

We have also seen some of the most common and intuitive notions of mathematics that we study at school: numbers, functions, equations – twisted, warped, sometimes even shattered. Many have proved to be nowhere near as fundamental as they had seemed. In modern math there are concepts and ideas that are deeper and more versatile: vector spaces, symmetry groups, arithmetic modulo prime numbers, sheaves. So mathematics has a lot more than meets the eye, and it is the Langlands Program that lets us begin to see what we could not see before. So far, we have only been able to catch glimpses of the hidden reality. And now, like archaeologists faced with a fractured mosaic, we try to piece together the evidence we were able to collect. Every new piece of the puzzle gives us new insights, new tools to unravel the mystery. And each time, we are dazzled by the seemingly inexhaustible richness of the emerging picture.

I found my own entry point into this magical world when Drinfeld connected my work on Kac–Moody algebras to the Langlands Program. This vast subject and its omnipresence in mathematics have fascinated me ever since. I became compelled to learn more and more about various tracks of the Program that are discussed in this book, and most of my research ever since has been either on the Langlands Program or was inspired by it in one way or another. This has forced me to travel across mathematical continents, learning different cultures and languages.

Like any traveler, I was bound to be surprised by what I saw. And now we come to one of the biggest surprises: it turns out that the Langlands Program is also inextricably linked to quantum physics. The key is duality, in physics as in math.

It might seem strange to look for a duality in physics, but in a sense this is a concept we are all already familiar with. Take electricity and magnetism. Even though these two forces seem to be quite different, they are actually described by a single mathematical theory, called electromagnetism. This theory possesses a hidden duality that exchanges electric and magnetic forces. (We will discuss it in detail below.) In the 1970s, physicists tried to generalize this duality to the so-called non-abelian gauge theories. These are the theories that describe nuclear forces: the "strong" force, which keeps quarks inside protons, neutrons, and other elementary particles; and the "weak" force, responsible for things like radioactive decay.

At the core of every gauge theory is a Lie group, which is called the *gauge group*. Electromagnetism is in a sense the simplest of gauge theories, and the gauge group is in this case our old friend, the circle group (the group of rotations of any round object). This group is abelian; that is, the multiplication of any two elements does not depend on the order in which the it is taken: $a \cdot b = b \cdot a$. But for the theories of strong and weak interactions, the corresponding gauge groups are non-abelian, that is, $a \cdot b \neq b \cdot a$ in the gauge group. And so we call them non-abelian gauge theories.

Now, in the 1970s, physicists found that there was an analogue of electromagnetic duality in the non-abelian gauge theories, but with a surprising twist. It turned out that if we start with the gauge theory whose gauge group is G, then the dual theory will be the gauge theory with another gauge group. And lo and behold, that group turned out to be nothing but the Langlands dual group ^LG, which is a key ingredient of the Langlands Program!

Think about it this way: mathematics and physics are like two different planets; say, Earth and Mars. On Earth, we discover a relation between different continents. Under this relation, every person in Europe gets matched with one in North America; their heights, weights, and ages are the same. But they have opposite genders (this is like switching a Lie group and its Langlands dual Lie group). Then one day we receive a visitor from Mars who tells us that on Mars they have also discovered a relation between their continents. Turns out every Martian on one of their continents can be matched with a Martian on another continent, so that their heights, weights, and ages are the same, but... they have opposite genders (who knew Martians had two genders, just like us?). We can't believe what we are hearing: it appears that the relation we have on Earth and the relation they have on Mars are somehow connected. But why?

Likewise, because the Langlands dual group appears in both math and physics, it is natural to assume that there must be a connection between the Langlands Program in mathematics and electromagnetic duality in physics. But for almost thirty years, no one could find it.

I discussed this question on several occasions over the years with Edward Witten. Professor at the Institute for Advanced Study in Princeton, he is considered as one of the greatest living theoretical physicists. One of his amazing qualities is the ability to use the most sophisticated apparatus of quantum physics to make astonishing discoveries and conjectures in pure mathematics. His work has inspired several generations of mathematicians, and he became the first physicist to earn the Fields Medal, one of the most prestigious prizes in mathematics.

Curious about a possible link between the quantum dualities and the Langlands Program, Witten would ask me about it from time to time. We would discuss it at my office at Harvard when he came to visit Harvard or MIT, or at his office in Princeton when I was there. The discussions were always stimulating, but we never got very far. It was clear that some essential elements were missing, still waiting to be discovered.

We got help from an unexpected source.

At a conference in Rome in May 2003,[1] I receive an e-mail from my old friend and colleague Kari Vilonen. Originally from Finland, Kari is one of the most gregarious mathematicians I know. When I first came to Harvard, he and his future wife Martina took me to a sports bar in Boston to watch a playoff baseball game in which the Red Sox were playing. Alas, the Sox lost, but what a memorable experience it was. We have been friends ever since, and years later we co-authored several papers on the Langlands Program (together with another mathematician, Dennis Gaitsgory). In particular, we proved together an important case of the Langlands relation.

In his e-mail, Kari (who was then a professor at Northwestern University) wrote that he had been contacted by people at DARPA who wanted to give us a grant to support research on the Langlands Program.

DARPA is the acronym for the Defense Advanced Research Projects Agency, the research arm of the U.S. Department of Defense. It was created in 1958 in the wake of the *Sputnik* launch with the mission to advance science and technology in the United States and to prevent the kind of technological surprise that *Sputnik* represented.

I read the following paragraph on the DARPA website:[2]

> To fulfill its mission, the Agency relies on diverse performers to apply multi-disciplinary approaches to both advance knowledge through basic research and create innovative technologies that address current practical problems through applied research. DARPA's scientific investigations span the gamut from laboratory efforts to the creation of full-scale technology demonstrations.... As the DoD's primary innovation engine, DARPA undertakes projects that are finite in duration but that create lasting revolutionary change.

Over the years, DARPA funded numerous projects in applied math and computer science; for example, it was responsible for the creation of ARPANET, the progenitor of the Internet. But as far as I knew, they had not supported projects in pure math. Why would they want to support research in the Langlands Program?

This area appeared to be pure and abstract, without immediate applications. But we have to realize that fundamental scientific research forms the basis of all technological progress. Often, what looked like the most abstract and abstruse discoveries in math and physics subsequently led to innovations that we now use in our everyday life. Think of the arithmetic modulo primes, for example. When we see it for the first time, it looks so abstract that it seems impossible something like this could have any real world applications. In fact, English mathematician G.H. Hardy famously argued that "great bulk of higher mathematics is useless."[3] But the joke was on him: many apparently esoteric results in number theory (his field of expertise) are now ubiquitous in, say, online banking. When we make online purchases, arithmetic modulo N springs into action (see the description of the RSA encryption algorithm in endnote 7 to Chapter 14). We should never try to prejudge the potential of a mathematical formula or idea for practical applications.

History shows that all spectacular technological breakthroughs were preceded, often decades earlier, by advances in pure research. Therefore, if we limit support for basic science, we limit our progress and power.

There is also another aspect of this: as a society, we are defined to a large extent by our scientific research and innovation. It is an important part of our culture and well-being. Robert Wilson, the first director of the Fermi National Laboratory, in which the largest particle accelerator of its era was created, put it this way in his testimony to the Congressional Joint Committee on Atomic Energy in 1969. Asked whether this multi-million dollar machine could help the country's security, he said:[4]

> Only from a long-range point of view, of a developing technology. Otherwise, it has to do with: Are we good painters, good sculptors, great poets? I mean all the things that we really venerate and honor in our country and are patriotic about. In that sense, this new knowledge has all to do with honor and country but it has nothing to do directly with defending our country except to make it worth defending.

Anthony Tether, who served as the director of DARPA from 2001 to 2009, recognized the importance of fundamental research. He challenged his program managers to find a good project in pure mathematics. One of the managers, Doug Cochran, took this call seriously. He had a friend at the National Science Foundation (NSF) by the name of Ben Mann. A specialist in the field of topology, Ben had left his academic position and come to Washington to serve as a program director at the Division of Mathematical Sciences of the NSF.

When Doug asked him to suggest a worthy project in pure mathematics, Ben thought of the Langlands Program. Even though this was not his area of expertise, he saw its importance from the grant proposals in this area submitted to the NSF. The quality of the projects and the fact that the same ideas propagated through diverse mathematical disciplines made a great impression on him.

So Ben suggested to Doug that DARPA support research in the Langlands Program, and that's why Kari, I, and two other mathematicians were contacted and asked to write a proposal that Doug would present to the director of DARPA. The expectation was that if the director approved it, we would receive a multi-million grant to direct research in this area.

In all honesty, we hesitated at first. This was an uncharted territory: no mathematicians we knew had ever received grants of this magnitude before. Normally, mathematicians receive relatively small individual grants from the NSF (a little travel money, support for a graduate student, and maybe some summer support). Here we would have to coordinate the work of dozens of mathematicians with the goal of making a concerted effort in a vast area of research. Because the grant would be so large, we would be subjected to much greater public scrutiny, and probably some measure of suspicion and jealousy from our colleagues. We recognized that without significant progress coming out of this project, we would be ridiculed, and that such a failure might close the door to funding other worthy projects in pure math by DARPA.

Despite our trepidation, we wanted to make an impact in the Langlands Program. And the idea of replacing the traditional, conservative scheme of funding mathematical research by a large injection of funds into a promising area sounded appealing and exciting. We simply couldn't say no.

The next question was what we should focus on in our project. The Langlands Program, as we have seen, is multi-faceted and relevant to many fields of mathematics. It would be easy to write half a dozen proposals on this general topic. We had to make a choice, and we decided to focus on what we thought was the biggest mystery: the potential link between the Langlands Program and dualities in quantum physics.

A week later, Doug made a presentation of our proposal to the DARPA director, and by all accounts, it was a success. The director approved a multi-million dollar funding for this project for three years. This was, as far as we could tell, the biggest grant awarded to research in pure mathematics to date. Expectations, obviously, were high. It was a moment of great excitement, but also some anxiety.

Fortunately for us, Ben Mann moved from the NSF to DARPA to be the program manager in charge of our project. From our first meetings with him, it was clear that Ben was uniquely qualified for this job. He has the vision and the courage to take on a high-risk/high-reward project, find the right people to implement it, and help them develop their ideas to the fullest. And his infectious enthusiasm energizes everybody around him. We were really lucky to have Ben at the helm. We would not have been able to achieve a fraction of what we did without his guidance and support.

As a first order of business, I sent an e-mail to Edward Witten telling him about our grant and asking him whether he would be interested in joining us. Given Witten's unique position in physics and math, we had to have him on board. Alas, Witten's first reaction was non-committal. He congratulated us on receiving the grant, but also made it clear that he had many projects to work on, and we shouldn't count on his participation.

But in a stroke of luck, Peter Goddard, one of the physicists who discovered the electromagnetic duality in non-abelian gauge theories, was about to become the director of the Institute for Advanced Study in Princeton. His more recent research was on things related to representation theory of Kac–Moody algebras, and because of this, I had met Peter at various conferences.

I remembered one of these meetings particularly well. It was in August 1991, and we were at a big workshop on math and quantum physics at Kyoto University in Japan. In the middle of the workshop, we received the alarming news of the *coup d'état* in the Soviet Union. It looked like the authoritarian regime was coming back to power, and the limited freedoms of *perestroika* would soon be scaled back. This would mean that the borders would be sealed again, and so there was a very real possibility that I wouldn't be able to see my family for years. My parents called me right away to tell me that if this were to happen, I should not worry about them and, in any event, I should not try to come back to Russia. As we said good-bye, we were preparing for the worst. It wasn't even clear that we would be able to talk to each other on the phone again in the near future.

Those were tumultuous days. One night, my good friend physicist Fedya Smirnov and I were in the lounge of one of the guest houses, watching Japanese TV and trying to figure out what was happening in Moscow. Everyone else in the building seemed to be sound asleep. Suddenly, around 3 am, Peter Goddard walked into the lounge, a bottle of Glenfiddich whiskey in hand. He asked us about the latest news, and we all had a drink. He then went back to sleep, but insisted that we keep the bottle – a nice gesture of support.

The next day the *coup d'état* was defeated, to our great relief. A picture of me and Borya Feigin (who was also at this conference) smiling and pumping fists in the air ended up on the front page of *Yomiuri*, one of the leading Japanese newspapers.

In my e-mail to Peter, I reminded him of this episode and told him about our DARPA grant. I suggested that we organize a meeting at the Institute for Advanced Study to bring together both physicists and mathematicians to talk about the Langlands Program and dualities in physics, to try to find common ground, so that we could solve the riddle together.

Peter's response was the best we could hope for. He offered his full support in organizing the meeting.

The institute was a perfect venue for such a meeting. Created in 1930 as an independent center of research and thinking, it has been home to Albert Einstein (who spent the last twenty years of his life there), André Weil, John von Neumann, Kurt Gödel, and other prominent scientists. The current faculty is equally impressive: it includes Robert Langlands himself, who has been a professor there since 1972 (now emeritus), and Edward Witten. Two other physicists on the faculty, Nathan Seiberg and Juan Maldacena, work in closely related areas of quantum physics, and several mathematicians, such as Pierre Deligne and Robert MacPherson, conduct research on topics linked to the Langlands Program.

My e-mail exchange with Goddard resulted in plans for an exploratory meeting in early December 2003. Ben Mann, Kari Vilonen, and I were coming to Princeton, and Goddard promised to participate. We invited Witten, Seiberg, and MacPherson; another Princeton mathematician, Mark Goresky, who was co-managing the DARPA project with Kari and myself, was to join us as well (we also invited Langlands, Maldacena, and Deligne, but they were traveling and could not attend).

The meeting was set to start at 11 am at the conference room next to the Institute cafeteria. Ben, Kari, and I arrived early, about fifteen minutes before the meeting. There was no one else there. As I was pacing nervously around the room, I couldn't stop thinking: "Is Witten coming?" He was the only one of the invitees who had not confirmed his participation.

Five minutes before the meeting, the door opened. It was Witten! That was the moment when I knew that something good would come out of all this.

A few minutes later, the other participants arrived. We all sat around a big table. After the usual greetings and small talk, there was silence; all eyes turned to me.

"Thank you all for coming," I began. "It has been known for some time that the Langlands Program and electromagnetic duality share something in common. But the exact understanding of what's going on has eluded us, despite numerous attempts. I think time has come to unravel this mystery. And now we have the necessary resources because we have received a generous grant from DARPA to support research in this area."

People at the table were nodding their heads. Peter Goddard asked, "How do you propose we go about it?"

Prior to the meeting, Kari, Ben, and I played out different scenarios, so I was well prepared.

"I suggest that we organize a meeting here at the institute. We will invite physicists working in related areas and we will organize lectures by mathematicians to present our current state of knowledge in the Langlands Program. We will then discuss together possible links to quantum physics."

Now all eyes turned to Witten, the dean of quantum physicists. His reaction was crucial.

Tall and physically imposing, Witten projects great intellectual power, to the point where some feel intimidated by him. When he speaks, his statements are precise and clear to a fault; they seem to be made of unbreakable logic. He never hesitates to take a pause, contemplating his answer. At such times, he often closes his eyes and leans his head forward. That's what he did at that moment.

All of us were waiting patiently. Less than a minute must have passed by, but to me it felt like eternity. Finally, Witten said, "This sounds like a good idea. What dates do you have in mind for the meeting?"

Ben, Kari, and I couldn't help but look at each other. Witten was on board, and this was a big victory for us.

After a brief discussion we found the dates that were suitable for everyone: March 8–10, 2004. Then someone asked who would be the participants and the speakers. We mentioned a few names and agreed to finalize the list over e-mail and send the invitations shortly. At this, the meeting adjourned. It took no more than fifteen minutes.

Needless to say, Ben, Kari, and I were very pleased. Witten promised to help organize the meeting (which would of course be a big draw for the invitees) and to actively participate in it as well. We also expected that Langlands would take part, as well as other physicists and mathematicians on the Institute faculty who were interested in the subject. Our first goal was accomplished.

In the course of the next few days we finalized the list of participants, and a week later invitations went out. The letter said:

> We are writing to invite you to participate in an informal workshop on the Langlands Program and Physics that will be held at the Institute for Advanced Study from March 8 to March 10, 2004. The goal of this workshop is to introduce physicists to recent developments in the geometric Langlands Program in the interest of exploring potential connections between this subject and Quantum Field Theory. We will plan several introductory lectures by mathematicians, and there will be ample time for informal discussions. This workshop is supported by a grant from DARPA.

Normally, conferences like this have fifty to a hundred participants. What often happens is that speakers give their talks while everyone listens politely. A couple of participants might ask questions at the end of the talk, and a few more may engage the speaker afterward. We envisioned something completely different: a dynamic event that was more a brainstorming session than a typical conference. Therefore we wanted to have a small meeting, about twenty people. We hoped that this format would encourage more interaction and freewheeling conversation between participants.

We already had our first meeting in this format in November 2003, at the University of Chicago. There was a small number of mathematicians invited, including Drinfeld and Beilinson (who had both taken professorships at the University of Chicago a few years earlier). That meeting was a success, and it proved to us that this format was working.

We decided that Kari, Mark Goresky, and I would speak, as well as my former Ph.D. student David Ben-Zvi, who was then a professor at the University of Texas at Austin. We broke down the material into four parts, each to be presented by one of us. In our presentations we had to convey the main ideas of the Langlands Program to the physicists who were not familiar with the subject. This was not an easy task.

Preparing for the conference, I wanted to learn more about the electromagnetic duality. We are all familiar with the electric and magnetic forces. Electric force is what makes electrically charged objects attract or repel each other depending on whether their charges are of the same or opposite signs. For example, an electron has negative electric charge, and a proton has a positive charge (of opposite value). The attractive force between them is what makes the electron spin around the nucleus of the atom. Electric forces create what is called an electric field. We have all seen it in action during a lightning strike, which is caused by the movement of warm wet air through an electric field.

Photo by Shane Lear. NOAA photo library.

Magnetic force has a different origin. It is the force that is created by magnets or by moving electrically charged particles. A magnet has two poles: north and south. When we place two magnets with opposite poles facing each other, they attract, whereas the same poles repel each other. The Earth is a giant magnet, and we take advantage of the magnetic force it exerts when we use a compass. Any magnet creates a magnetic field, as we can see clearly on the picture on the next page.

In the 1860s, British physicist James Clerk Maxwell developed an exquisite mathematical theory of electric and magnetic fields. He described them by a system of differential equations that now carry his name. You might expect these equations to be long and complex, but in fact they are quite simple: there are only four of them, and they look surprisingly symmetrical. It turns out that if we consider the theory in the vacuum (that is, without any matter present), and exchange the electric field and magnetic fields, the system of equations will not change.[5] In other words, the switching of the two fields is a symmetry of the equations. It is called the electromagnetic duality. This means the relationship between the electric and magnetic fields is symmetrical: each of them affects the other in exactly the same way.

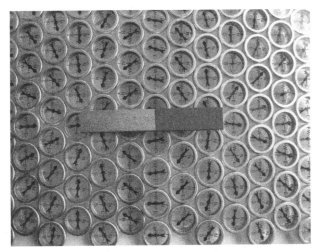

Photo by Dayna Mason.[6]

Now, Maxwell's beautiful equations describe *classical* electromagnetism, in the sense that this theory works well at large distances and low energies. But at small distances and high energies, the behavior of the two fields is described by the *quantum* theory of electromagnetism. In the quantum theory, these fields are carried by elementary particles, photons, which interact with other particles. This theory goes under the name of quantum field theory.

To avoid confusion, I want to stress that the term "quantum field theory" has two different connotations: in a broad sense, it means the general mathematical language that is used to describe the behavior and interaction of elementary particles; but it may also refer to a particular model of such behavior – for example, quantum electromagnetism is a quantum field theory in this sense. I will mostly use the term in the latter sense.

In any such theory (or model), some particles (like electrons and quarks) are the building blocks of matter, and some (like photons) are the conduits of forces. Each particle has various characteristics: some familiar ones, like mass and electric charge, and some less familiar, like "spin." A particular quantum field theory is then a recipe to combine them together.

Actually, the word "recipe" points us toward a useful analogy: think of a quantum field theory as a culinary recipe. Then the ingredients of the dish we are cooking are the analogues of particles, and the way we mix them together is like the interaction between the particles.

For example, let's look at this recipe of the Russian soup *borscht*, a perennial favorite in my home country. My mom makes the best one (of course!). Here's what it looks like (the picture was taken by my dad):

Obviously, I have to keep my mom's recipe secret. But here's a recipe I found online:

8 cups of broth (beef or vegetable)
1 pound slice of bone-in beef shank
1 large onion
4 large beets, peeled
4 carrots, peeled
1 large russet potato, peeled
2 cups of sliced cabbage
3/4 cup of chopped fresh dill
3 table spoon of red wine vinegar
1 cup sour cream
Salt
Pepper

Think of this as the "particle content" of our quantum field theory. What would the duality mean in this context? It would simply mean exchanging some of the ingredients ("particles") with others, so that the total content stays the same.

Here is how such a duality could work:

beet $\longrightarrow$ carrot
carrot $\longrightarrow$ beet
onion $\longrightarrow$ potato
potato $\longrightarrow$ onion
salt $\longrightarrow$ pepper
pepper $\longrightarrow$ salt

All other ingredients stay put under the duality; that is,

broth $\longrightarrow$ broth
beef shank $\longrightarrow$ beef shank

and so on.

Since the amounts of the ingredients we exchange are the same, the result will be the same recipe! This is the meaning of duality.

If, on the other hand, we exchanged beets for potatoes, we would get a different recipe: one that would have four potatoes and only one beet. I haven't tried it, but I am guessing it would taste awful.

It should be clear from this example that a symmetry of a recipe is a rare property, from which we can learn something about the dish. The fact that we can switch beets with carrots without affecting the outcome means that our borscht is well-balanced between them.

Let's go back to quantum electromagnetism. Saying that there is a duality in this theory means that there is a way to exchange the particles so that we end up with the same theory. Under the electromagnetic duality we want all "things electric" to become "things magnetic," and vice versa. So, for instance, an electron (an analogue of a beet in our soup) carries an electric charge, so it should be exchanged with a particle that carries a magnetic charge (an analogue of a carrot).

The existence of such a particle contradicts our everyday experience: a magnet always has two poles, and they cannot be separated! If we break a magnet in two pieces, each of them will also have two poles.

Nonetheless, the existence of a magnetically charged elementary particle, called magnetic monopole, has been theorized by physicists; the first was one of the founders of quantum physics, Paul Dirac, in 1931. He showed that if we allow something funny to happen to the magnetic field at the position of the monopole (this is what a mathematician would call a "singularity" of the magnetic field), then it will carry magnetic charge.

Alas, magnetic monopoles have not been discovered experimentally, so we don't know yet whether they exist in nature. If they don't exist, then an exact electromagnetic duality does not exist in nature at the quantum level.

The jury is still out on whether this is the case or not. Regardless, we can try to build a quantum field theory that is close enough to nature and exhibits the electromagnetic duality. Going back to our kitchen analogy, we can try to "cook up" new theories that possess dualities. We can change the ingredients and their quantities in recipes we know, get rid of some of them, throw in something extra, and so on. This kind of "experimental cuisine" may lead to variable results. We may not necessarily want to "eat" these imagined dishes. But edible or not, it may be worthwhile to study their properties in our dreamed-up kitchen – they may give us some clues about the dishes that are edible (that is to say, the models that could describe our universe).

This trial-and-error "model building" is a path along which progress has been made in quantum physics for decades (just as it was in the culinary art). And symmetry is a powerful guiding principle that has been used in creating these models. The more symmetrical a model is, the easier it is to analyze.

At this point, it is important to note that there are two kinds of elementary particles: fermions and bosons. The former are the building blocks of matter (electrons, quarks, etc.), and the latter are the particles that carry forces (such as photons). The elusive Higgs particle, discovered recently at the Large Hadron Collider under Geneva, is also a boson.

There is a fundamental difference between the two types of particles: two fermions cannot be in the same "state" simultaneously, whereas any number of bosons can. Because their behavior is so radically different, for a long time physicists assumed that any symmetry of a quantum field theory had to preserve a distinction between the fermionic and bosonic sectors – that nature forbids them to be mixed together. But in the mid-1970s several physicists suggested what looked like a crazy idea: that a new type of symmetry was possible that would exchange bosons with fermions. It was christened *supersymmetry*.

As Niels Bohr, one of the creators of quantum mechanics, famously said to Wolfgang Pauli, "We are all agreed that your theory is crazy. The question that divides us is whether it is crazy enough to have a chance of being correct."

In the case of supersymmetry, we still don't know whether it is realized in nature, but the idea has become popular. The reason is that many of the issues that plague conventional quantum field theories are eliminated when supersymmetry is introduced. Supersymmetric theories are generally more elegant and easier to analyze.

Quantum electromagnetism is not supersymmetric, but it has supersymmetric extensions. We throw in more particles, both bosons and fermions, so that the resulting theory exhibits supersymmetry.

In particular, physicists have studied the extension of the electromagnetism with the maximal possible amount of supersymmetry. And they showed that in this extended theory the electromagnetic duality is indeed realized.

To summarize, we don't know whether a form of quantum electromagnetic duality exists in the real world. But we do know that in an idealized, supersymmetric, extension of the theory, the electromagnetic duality is manifest.

There is one important aspect of this duality that we haven't yet discussed. The quantum field theory of electromagnetism has a parameter: the electric charge of the electron. It is negative, so we write it as $-e$, where $e = 1.602 \cdot 10^{-19}$ Coulombs. It is very small. The maximal supersymmetric extension of electromagnetism has a similar parameter, which we will also denote by e. If we perform the electromagnetic duality and exchange all things electric by all things magnetic, we will get a theory in which the charge of the electron will be not e, but its inverse, $1/e$.

If e is small, then $1/e$ is large. So if we start with the theory with a small charge of the electron (as is the case in our world), then the dual theory will have a large charge of the electron.

This is hugely surprising! In terms of our soup analogy, imagine that e is the soup temperature. Then the duality would mean that switching the ingredients such as carrots and beets would suddenly convert a cold borscht into a hot one.

This inversion of e is in fact a key aspect of the electromagnetic duality, which has far-reaching consequences. The way quantum field theory is set up, we have a good handle on the theory only for small values of the parameter such as e. We don't even know *a priori* that the theory makes sense at large values of the parameter. Electromagnetic duality tells us not only that it makes sense, but that it is in fact equivalent to the theory with small values of the parameter. This means that we have a chance to describe the theory for all values of the parameter. That's why this kind of duality is considered as a Holy Grail of quantum physics.

Our next question is whether electromagnetic duality exists for quantum field theories other than the electromagnetism and its supersymmetric extension.

Apart from the electric and magnetic forces, there are also three other known forces of nature: gravity, which we all know and appreciate, and the two nuclear forces with rather mundane names: strong and weak. The strong nuclear force is keeping quarks inside elementary particles such as protons and neutrons. The weak nuclear force is responsible for various processes transforming atoms and elementary particles, such as the so-called beta-decay of atoms (emission of electrons or neutrinos) and hydrogen fusion, which powers stars.

These forces seem to be quite distinct. It turns out, however, that the theories of electromagnetic, weak, and strong forces have something in common: they are what we call gauge theories, or Yang–Mills theories, in honor of the physicists Chen Ning Yang and Robert Mills, who wrote a groundbreaking paper about them in 1954. As I mentioned at the beginning of this chapter, gauge theory has a symmetry group, called gauge group. It is a Lie group, a concept we talked about in Chapter 10. The gauge group of the theory of electromagnetism is the group that I introduced at the very beginning of this book, the circle group (also known as $SO(2)$ or $U(1)$). It is the simplest Lie group, and it is abelian. We already know that many Lie groups are non-abelian, such as the group $SO(3)$ of rotations of a sphere. The idea of Yang and Mills was to construct a generalization of the electromagnetism in which the circle group would be replaced by a non-abelian group. It turned out that gauge theories with non-abelian gauge groups accurately describe the weak and strong nuclear forces.

The gauge group of the theory of the weak force is the group called $SU(2)$. It is the Langlands dual group to $SO(3)$ and is twice as big (we talked about it in Chapter 15). The gauge group of strong nuclear force is called $SU(3)$.[7]

So gauge theories provide a universal formalism describing three out of four fundamental forces of nature (we count electric and magnetic forces as parts of one force of electromagnetism). Moreover, in subsequent years, it was realized that these are not just three separate theories but parts of a whole: there is one theory, commonly referred to as the Standard Model, which includes the three forces as different pieces. It is therefore what we could call a "unified theory" – something Einstein sought unsuccessfully during the last thirty years of his life (though at that time only two forces were known: electromagnetism and gravity).

We have already talked at length about the importance of the idea of unification in mathematics. For example, the Langlands Program is a unified theory in the sense that it describes a wide range of phenomena in different mathematical disciplines in similar terms. The idea of building a unified theory from as few first principles as possible is especially appealing in physics, and it's clear why. We would like to reach the most complete understanding of the inner workings of the universe, and we hope that the ultimate theory – if it exists – is simple and elegant.

Simple and elegant do not mean easy. For example, Maxwell's equations are deep, and it takes one some effort to understand what they mean. But they are simple in the sense that they are most economical in expressing the truth about the electric and magnetic forces. They are also elegant. So are Einstein's equations of gravity and the equations of non-abelian gauge theory found by Yang and Mills. A unified theory should combine all of them, like a symphony weaving together the voices of different instruments.

The Standard Model is a step in this direction, and its experimental confirmation (including the recent discovery of the Higgs boson) was a triumph. However, it is not the ultimate theory of the universe: for one thing, it does not include the force of gravity, which has proved to be the most elusive one. Einstein's general relativity theory gives us a good understanding of gravity classically, that is, at large distances, but we still don't have an experimentally testable quantum theory that would describe the force of gravity at very short distances. Even if we focus on the three other forces of nature, the Standard Model still leaves too many questions unanswered and does not account for a big chunk of matter observed by astronomers (called "dark matter"). So the Standard Model is but a partial draft of the ultimate symphony.

One thing is clear: the final score of the ultimate symphony will be written in the language of mathematics. In fact, after Yang and Mills wrote their celebrated paper introducing non-abelian gauge theories, physicists realized, to their astonishment, that the mathematical formalism needed for these theories was developed by mathematicians decades earlier, without any reference to physics. Yang, who went on to win a Nobel Prize, described his awe in these words:[8]

> [I]t was not just joy. There was something more, something deeper: After all, what could be more mysterious, what could be more awe-inspiring, than to find that the structure of the physical world is intimately tied to the deep mathematical concepts, concepts which were developed out of considerations rooted only in logic and the beauty of form?

The same sense of wonder was expressed by Albert Einstein when he asked,[9] "How can it be that mathematics, being after all a product of human thought independent of experience, is so admirably appropriate to the objects of reality?"

The concepts that Yang and Mills used to describe forces of nature appeared in mathematics earlier because they were natural also within the paradigm of geometry that mathematicians were developing following the inner logic of that subject. This is a great example of what another Nobel Prize–winner, physicist Eugene Wigner, called the "unreasonable effectiveness of mathematics in the natural sciences."[10] Though scientists have been exploiting this "effectiveness" for centuries, its roots are still poorly understood. Mathematical truths seem to exist objectively and independently of both the physical world and the human brain. There is no doubt that the links between the world of mathematical ideas, physical reality, and consciousness are profound and need to be further explored. (We will talk more about this in Chapter 18.)

We also need new ideas in order to go beyond the Standard Model. One such idea is supersymmetry. Whether it is present in our universe is the subject of a big debate. So far, no traces of it have been discovered. The experiment is the ultimate judge of a theory, so until proved experimentally, supersymmetry remains a theoretical construct, no matter how beautiful and alluring the idea may be. But even if it turns out that supersymmetry is not realized in the real world, it provides a convenient mathematical apparatus that we can use for building new models of quantum physics. These models are not that far away from the models governing the physics of the real world but are often much easier to analyze because of the greater degree of symmetry that they exhibit. We hope that what we learn about these theories will have a bearing on the realistic theories of our universe, regardless of whether supersymmetry exists in it or not.

Just like the theory of electromagnetism has a maximal supersymmetric extension, so do non-abelian gauge theories. These supersymmetric theories are obtained by throwing in more particles into the mix, both bosons and fermions, to reach the most perfect possible balance between them. It is then natural to ask: do these theories possess an analogue of the electromagnetic duality?

Physicists Claus Montonen and David Olive tackled this question[11] in the late 1970s. Building on an earlier work[12] by Peter Goddard (the future director of the Institute for Advanced Study), Jean Nuyts, and David Olive, they came up with a startling conclusion: yes, there is an electromagnetic duality in the supersymmetric non-abelian gauge theories, but these theories are not self-dual in general, the way electromagnetism is. As we discussed above, if we replace all things electric by all things magnetic and vice versa in electromagnetism, we will get back the same theory, with the inverted charge of electron.

But it turns out that if we do the same in a general supersymmetric gauge theory with a gauge group G, we will obtain a *different* theory. It will still be a gauge theory, but with a different gauge group (and also with the inverted parameter, which is the analogue of the charge of the electron).

And what will the gauge group be in the dual theory? It turns out to be ^LG, the Langlands dual group of the group G.

Goddard, Nuyts, and Olive discovered it by performing a detailed analysis of the electric and magnetic charges in the gauge theory with a gauge group G. In electromagnetism, which is the gauge theory with the gauge group being the circle group, the values of both charges are integers. When we exchange them, one set of integers gets exchanged with another set of integers. Hence the theory stays the same. But they showed that, in a general gauge theory, the electric and magnetic charges take values in two different sets. Let's call them S_e and S_m. They can be expressed mathematically in terms of the gauge group G (it is not important at the moment how, exactly).[13]

It turns out that under the electromagnetic duality, S_e becomes S_m and S_m becomes S_e. So the question is whether there is another group G', for which S_e is what was S_m for G, and S_m is what was S_e for G (this should also be compatible with some additional data determined by G and G'). It is not obvious whether such a group G' exists, but they showed that it does and gave a construction. They didn't know at the time that G' had already been constructed by Langlands a decade earlier in much the same way, even though Langlands' motivation was entirely different. This group G' is nothing but the Langlands dual group ^LG.

Why the electromagnetic duality leads to the same Langlands dual group that mathematicians discovered in a totally different context was the big question we were going to tackle at the meeting in Princeton.

Chapter 17

Uncovering Hidden Connections

Just about an hour by train from New York City, Princeton looks like a typical Northeastern suburban town. The Institute for Advanced Study, known in the scientific community simply as *"the* institute," is on the outskirts of Princeton, literally in the woods. The area around it is quiet and picturesque: ducks swimming in small ponds, trees reflected in still water. The institute, a cluster of two- and three-story brick buildings with the feel of the 1950s, radiates intellectual power. One can't help but savor its rich history wandering in the hushed corridors and the main library, which was used by Einstein and other giants.

This is where we had our meeting in March 2004. Despite the short notice, the response to the invitations we sent in December was overwhelmingly positive. There were about twenty participants – so when I opened the meeting, I asked those present to take turns and introduce themselves. I felt like pinching myself: Witten and Langlands were there, sitting close by, as was Peter Goddard – and several of their colleagues from both School of Mathematics and School of Natural Sciences. David Olive, of the Montonen–Olive and Goddard–Nuyts–Olive papers, was also present. And of course Ben Mann was with us as well.

Everything went according to plan. We were essentially recounting the story that you have been reading in this book: the origins of the Langlands Program in number theory and harmonic analysis, the passage to curves over finite fields and then to the Riemann surfaces. We also spent quite a bit of time explaining the Beilinson–Drinfeld construction and my work with Feigin on Kac–Moody algebras, as well as its links to the two-dimensional quantum field theories.

Unlike a typical conference, there was a lot of give and take between the speakers and the audience. It was an intense meeting, with discussions continuing from seminar room to cafeteria and back.

Throughout, Witten was in high gear. He was sitting in the front row, listening intently and asking questions, constantly engaging the speakers. On the morning of the third day, he said to me, "I'd like to speak in the afternoon; I think I have an idea what's going on."

After lunch, he gave an outline of a possible connection between the two subjects. This was the beginning of a new theory bridging math and physics, which he and his collaborators, and then many others, have been pursuing ever since.

As we discussed, in the third column of André Weil's Rosetta stone, the geometric version of the Langlands Program revolves around Riemann surfaces. All of these surfaces are two-dimensional. For example, as we discussed in Chapter 10, the sphere – the simplest Riemann surface – has two coordinates: latitude and longitude. That's why it is two-dimensional. All other Riemann surfaces are two-dimensional as well because a small neighborhood of each point looks like a piece of a two-dimensional plane, so it may be described by two independent coordinates.

On the other hand, gauge theories, in which the electromagnetic duality is observed, are defined in the four-dimensional space-time. In order to bridge the two, Witten started by applying a "dimensional reduction" of a four-dimensional gauge theory from four to two dimensions.

Dimensional reduction is actually a standard tool in physics: we approximate a given physical model by focusing on some degrees of freedom while ignoring others. For example, suppose you are flying on a plane, and a flight attendant, standing in the aisle, gives you a glass of water. Assume for simplicity that the motion of the flight attendant's hand is perpendicular to the motion of the plane. The velocity of the glass has two components: the first is the velocity of the plane, and the second is the velocity of the flight attendant's hand, passing you the glass. But the former is much larger than the latter, so if we were to describe the motion of the glass in the air from the point of view of a static observer on the ground, we could safely ignore the second component of velocity and simply say that the glass is moving with the same velocity as the plane. Therefore, we can reduce a two-dimensional problem involving two components of velocity to a one-dimensional problem involving the component that dominates.

In our context, the dimensional reduction is realized as follows: we consider a geometric shape (or manifold), which is the product of two Riemann surfaces. Here, "product" means that we consider a new geometric shape whose coordinates are the coordinates of each of these surfaces, taken together.

As a simpler example, let's consider the product of two lines. Each line has one coordinate, so the product will have two independent coordinates. Hence it will be a plane: each point on the plane is represented by a pair of coordinates. These are the coordinates of the two lines, taken together.

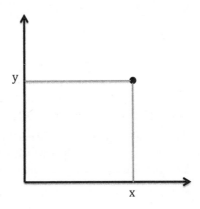

Likewise, the product of a line and a circle is a cylinder. It also has two coordinates; one circular and one linear.

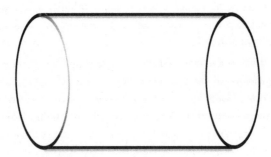

When we take the product, the dimensions add up. In the examples we have just considered, each of the initial two objects is one-dimensional, and their product is two-dimensional. Here is another example: the product of a line and a plane is the three-dimensional space. Its dimension is $3 = 1 + 2$.

Likewise, the dimension of the product of two Riemann surfaces is the sum of their dimensions, that is $2 + 2$, which is 4. We can draw a picture of a Riemann surface (we have seen some of them earlier), but we can't draw a four-dimensional manifold, so we will just have to study it mathematically, using the same methods that we use for the lower-dimensional shapes, which we can imagine more easily. Our ability to do so is a good illustration of the power of mathematical abstraction, as we have already discussed in Chapter 10.

Now suppose that the size of one of the two Riemann surfaces – call it X – is much smaller than the size of the other one, which we will call Σ. Then the effective degrees of freedom will be concentrated on Σ and we will be able to approximately describe the four-dimensional theory on the product of the two surfaces by a theory on Σ, which physicists refer to as an "effective theory." This theory will then be two-dimensional. This approximation will become better and better as we rescale X to make its size smaller and smaller, while preserving its shape (note that this effective theory will still depend on the shape of X). Thus, we pass from the four-dimensional supersymmetric gauge theory on the product of X and Σ to a two-dimensional theory defined on Σ.

Before we discuss the nature of this theory in any detail, let's talk about what we mean by a quantum field theory in general. For example, in electromagnetism, we study electric and magnetic fields in three-dimensional space. Each of them is what mathematicians call a vector field. A useful analogy is the vector field describing a wind pattern: at each point in space, the wind blows in a particular direction and has a particular strength – and this is captured by a pointed line segment attached to this point, which mathematicians call a vector. The collection of these vectors attached to all points in space is a vector field. We have all seen the wind represented as a vector field on weather maps.

Likewise, a given magnetic field has a particular direction and strength at each point in space, as can be seen from the picture on p. 195. Hence, it is also a vector field. In other words, we have a rule that assigns a vector to each point of our three-dimensional space. Not surprisingly, mathematicians call such a rule "a map" from our three-dimensional space to the three-dimensional vector space. And if we follow how a given magnetic field changes in time, we obtain a map from the four-dimensional space-time to the three-dimensional vector space. (This is like watching how the weather map is changing over time on TV.) Similarly, any given electric field, changing in time, may also be described as a map from the four-dimensional space-time to the three-dimensional vector space. Electromagnetism is a mathematical theory describing these two maps.

In the classical theory of electromagnetism, the only maps that we are interested in are the maps corresponding to solutions of the Maxwell equations. In contrast, in quantum theory we study *all* maps. Any calculation in quantum field theory, in fact, involves the summation over all possible maps, but each map is weighted, that is, multiplied by a prescribed factor. These factors are defined in such a way that the maps corresponding to solutions of the Maxwell equations make the dominant contribution, but other maps also contribute.

Maps from space-time to various vector spaces appear in many other quantum field theories (for example, in non-abelian gauge theories). However, not all quantum field theories rely on vectors. There is a class of quantum field theories, called the *sigma models*, in which we consider maps from the space-time to a curved geometric shape, or manifold. This manifold is called the target manifold. For example, it could be a sphere. Though sigma models were first studied in the case of four-dimensional space-time, such a model also makes sense if we take space-time to be a manifold of any dimension. Thus, there is a sigma model for any choice of the target manifold and any choice of the space-time manifold. For example, we can choose a two-dimensional Riemann surface as our space-time, and the Lie group $SO(3)$ as the target manifold. Then the corresponding sigma model will describe maps from this Riemann surface to $SO(3)$.

The picture below illustrates such a map: on the left-hand side we have a Riemann surface, on the right-hand side we have the target manifold, and the arrow represents a map between them; that is, a rule that assigns a point in the target manifold to each point in the Riemann surface.

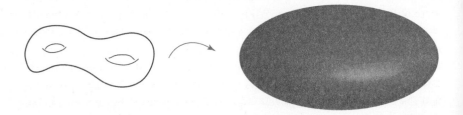

In the classical sigma model, we consider maps from the space-time to the target manifold that solve the equations of motion (the analogues of the Maxwell equations of electromagnetism); such maps are called harmonic. In the quantum sigma model, all quantities of interest, such as the so-called correlation functions, are computed by summing up over all possible maps, with each map weighted, that is, multiplied by a prescribed factor.

Let's return to our question: which two-dimensional quantum field theory describes the dimensional reduction of a four-dimensional supersymmetric gauge theory with the gauge group G on $\Sigma \times X$ as we rescale X so that its size becomes very small? It turns out that this theory is a supersymmetric extension of the sigma model of maps from Σ to a specific target manifold M, which is determined by the Riemann surface X and the gauge group G of the original gauge theory. Our notation for it should reflect this, so we will denote it by $M(X,G)$.[1]

As had earlier turned out to be the case with group theory (see Chapter 2), when physicists stumbled upon these manifolds, they discovered that mathematicians had been there before them. In fact, these manifolds had a name: *Hitchin moduli spaces*, after British mathematician Nigel Hitchin, professor at Oxford University, who had introduced and studied these spaces in the mid-1980s. Although it is clear why a physicist would be interested in these spaces – they appear when we do the dimensional reduction of a four-dimensional gauge theory – the reasons for a mathematician's interest in these spaces seem less obvious.

Luckily, Nigel Hitchin has given a detailed account[2] of the history of his discovery, and it's actually a great example of the subtle interplay between math and physics. In the late 1970s, Hitchin, Drinfeld, and two other mathematicians, Michael Atiyah and Yuri Manin, studied the so-called instanton equations, which physicists had come up with while studying gauge theories. These instanton equations were written in a flat four-dimensional space-time. Hitchin subsequently studied differential equations in a flat three-dimensional space, called the monopole equations, obtained by dimensional reduction of the instanton equations from four to three dimensions. Those were interesting from a physical point of view, and they also turned out to have an intriguing mathematical structure.

It was then natural to look at the differential equations obtained by reducing the instanton equations from four to two dimensions. Alas, physicists had observed that these equations did not have any non-trivial solutions in the flat two-dimensional space (that is, on the plane), so they did not pursue these equations further. Hitchin's insight, however, was that these equations could be written on any *curved* Riemann surface, such as the surface of a donut or a pretzel, as well. Physicists missed this, because at the time (in the early 1980s) they were not particularly interested in quantum field theories on such curved surfaces. But Hitchin saw that, mathematically, solutions on these surfaces were quite rich. He introduced his moduli space $M(X,G)$ as the space of solutions of these equations on a Riemann surface X (in the case of a gauge group G).* He found that it was a remarkable manifold; in particular, it possessed a "hyper-Kähler metric," of which very few examples were known at the time. Other mathematicians followed in his footsteps.

About ten years later, physicists began to appreciate the importance of these manifolds in quantum physics, although that interest did not really catch on before the work of Witten and his collaborators, which I am presently describing. (It is also interesting to note that the Hitchin moduli spaces, which originally appeared in the right column of André Weil's Rosetta stone, recently found applications to the Langlands Program in the middle column, in which the role of Riemann surfaces is played by curves over finite fields.[3])

The interaction between math and physics is a two-way process, with each of the two subjects drawing from and inspiring the other. At different times, one of them may take the lead in developing a particular idea, only to yield to the other subject as focus shifts. But altogether, the two interact in a virtuous circle of mutual influence.

Now, armed as we are with the insights of both mathematicians and physicists, let's apply the electromagnetic duality to the four-dimensional gauge theory with the gauge group G. We will then obtain the gauge theory with the gauge group ^LG, the Langlands dual group of G. (Recall that if we apply this duality twice, we get back the original group G. In other words, the Langlands dual group of ^LG is the group G itself.) The effective two-dimensional sigma models on Σ, associated to G and ^LG, will then also be equivalent, or dual, to each other. For the sigma models this kind of duality is called *mirror symmetry*. In one of the sigma models we consider maps from Σ to the Hitchin moduli space $M(X,G)$ corresponding to G; in the other we consider maps from Σ to the Hitchin moduli space $M(X,{}^LG)$ corresponding to ^LG. The two Hitchin moduli spaces, and their sigma models, have nothing to do with each other *a priori*, so

---

*In this regard, Hitchin quotes the great German poet Goethe: "Mathematicians are like Frenchmen: whatever you say to them they translate into their own language, and forthwith it is something entirely different."

the mirror symmetry between them is just as surprising as the electromagnetic duality of the original gauge theories in four dimensions.

Physicists' interest in two-dimensional sigma models of this type is motivated in part by the important role they play in string theory. As I mentioned in Chapter 10, string theory postulates that fundamental objects of nature are not point-like elementary particles (which have no internal geometry and hence are zero-dimensional), but are one-dimensional extended objects called strings, which can be open or closed. The former have two end-points, and the latter are little loops, much like the ones we encountered in Chapter 10.

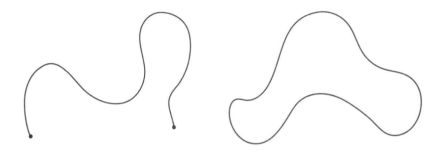

The idea of string theory is that vibrations of these tiny strings as they move through space-time create elementary particles and forces between them.

Sigma models enter string theory once we begin to consider how strings move. In standard physics, when a point-like particle moves in space, its trajectory is a one-dimensional path. The positions of the particle at different moments in time are represented by points on this path.

If a closed string is moving, however, then its movement sweeps a two-dimensional surface. Now the position of the string at a particular moment in time is a loop on this surface.

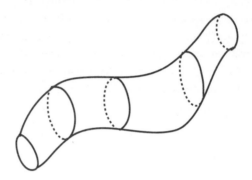

The strings may also interact with each other: a string may "split" into two or more pieces, and those pieces may also come together, as shown on the next picture. This gives us a more general Riemann surface with an arbitrary number of "holes" (and with boundary circles). It is called the worldsheet of the string.

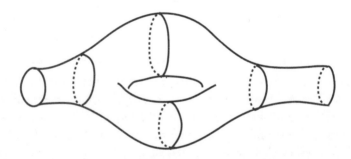

Such a trajectory may be represented by a Riemann surface Σ embedded into space-time S and hence by a map from Σ to S. These are precisely the kinds of maps that appear in the sigma model on Σ with the target manifold S. However, things are now turned upside-down: the space-time S is now the target manifold of this sigma model – that is, the recipient of the maps – not the source of the maps, in contrast to the conventional quantum field theories, such as electromagnetism.

The idea of string theory is that by doing calculations in these sigma models *and* summing up the results over all possible Riemann surfaces Σ (that is, over all possible paths of the strings propagating in a fixed space-time $S)^4$ we can reproduce the physics that we observe in space-time S.

Unfortunately, the resulting theory is plagued by some serious problems (in particular, it allows for the existence of "tachyons," elementary particles moving faster than light, whose existence is prohibited by Einstein's relativity theory). The situation improves dramatically if we consider a supersymmetric extension of string theory. Then we obtain what's called *superstring theory*. But again, there is a problem: superstring theory turns out to be mathematically consistent only if our space-time S has ten dimensions, which is at odds with the world we observe having only four (three dimensions of space and one time dimension).

However, it could be that our world is really a product, in the sense explained above, of the four-dimensional space-time we observe and a tiny six-dimensional manifold M, which is so small that we cannot see it using available tools. If so, then we would be in a situation similar to the dimensional reduction (from four to two dimensions) that we have discussed above: the ten-dimensional theory would give rise to an effective four-dimensional theory. The hope is that this effective theory describes our universe, and in particular, includes the Standard Model as well as a quantum theory of gravity. This promise of potential unification of all known forces of nature is the main reason that superstring theory has been studied so extensively in recent years.[5]

But we have a problem: which six-dimensional manifold is this M?

To appreciate just how daunting this problem is, let's suppose for the sake of argument that superstring theory were mathematically consistent in six dimensions rather than ten. Then there would only be two extra dimensions and we would have to find a two-dimensional manifold M. In this case, there wouldn't be so many choices: M would have to be a Riemann surface, which, as we know, is characterized by its genus, that is, the number of "holes." Furthermore, it turns out that for the theory to work, this M has to satisfy certain additional properties; for example, it has to be what's called a Calabi–Yau manifold, in honor of two mathematicians, Eugenio Calabi and Shing-Tung Yau, who were the first to study such spaces mathematically (years before physicists got interested in the subject, I might add).[6] The only Riemann surface that has this property is the torus. Hence, if M were two-dimensional, we would be able to pin it down – it would have to be a torus.[7] However, as the dimension of

M grows, so does the number of possibilities. If M is six-dimensional, then by some estimates there are 10^{500} choices – an unimaginably large number. Which one of these six-dimensional manifolds is realized in our universe and how can we verify this experimentally? This is one of the key questions of string theory that still remains unanswered.[8]

In any case, it should be clear from this discussion that sigma models play a crucial role in superstring theory, and in fact their mirror symmetry may be traced to a duality in superstring theory.[9] Sigma models also have many applications outside of string theory. Physicists have been studying them in great detail, and not only the sigma models in which the target manifold M is six-dimensional.[10]

So, when Witten spoke at our conference in 2004, he first applied the technique of dimensional reduction (from four to two dimensions), to reduce the electromagnetic duality of two gauge theories (with gauge groups G and ^LG) to the mirror symmetry of two sigma models (with the targets being the Hitchin moduli spaces associated to the two Langlands dual groups, G and ^LG). Then he asked: can we connect this mirror symmetry to the Langlands Program?

The answer he suggested was fascinating. Usually, in quantum field theories, we study something called correlation functions, which describe the interaction of particles; for example, one such function might be used to describe the probability of a certain particle emerging from the collision of two others. But it turns out that the formalism of quantum field theory is much more versatile: in addition to these functions, there are also various more subtle objects present in the theory, which are similar to the "sheaves" that we discussed in Chapter 14 in connection with Grothendieck's dictionary. These objects go under the name "D-branes," or simply "branes."

Branes originated in superstring theory, and their name is a truncation of the word "membrane." Branes arise naturally when we consider the movement of open strings on a target manifold M. The simplest way to describe the positions of both ends of an open string is to stipulate that one of the end points belongs to a particular subset, B_1, of M and the other belongs to another subset, B_2. This is shown on the picture below, on which the thin curve represents the open string with two end points, one of which is on B_1 and the other on B_2.

This way, the subsets (or, more properly, submanifolds) B_1 and B_2 become players in superstring theory and in the corresponding sigma model. These subsets are the prototypes of the general branes that occur in these theories.[11]

The mirror symmetry between two sigma models gives rise to a relation between the branes in these two sigma models. The existence of this relation was originally proposed in the mid-1990s by mathematician Maxim Kontsevich under the name "homological mirror symmetry." It has been extensively studied by both physicists and mathematicians, especially in the last decade.

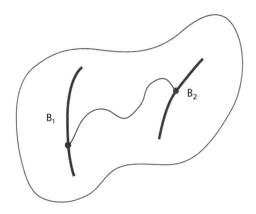

The main idea of Witten's talk in Princeton was that it is this homological mirror symmetry that should be equivalent to the Langlands relation.

At this point, it is important to note that sigma models come in two flavors, called the "A-model" and the "B-model." The two sigma models we are considering are in fact different: if the one with the target manifold Hitchin moduli space $M(X,G)$ is the A-model, then the one with the target manifold $M(X,{}^L G)$ is the B-model. Accordingly, the branes in the two theories are called "A-branes" and "B-branes," respectively. Under the mirror symmetry, for each A-brane on $M(X,G)$ there should be a B-brane on $M(X,{}^L G)$, and vice versa.[12]

In order to establish the geometric Langlands relation, we need to associate an automorphic sheaf to each representation of the fundamental group of X in ${}^L G$. Here is roughly how Witten proposed it could be constructed using mirror symmetry:

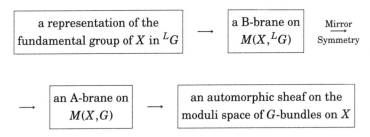

Though many details remained to be ironed out, Witten's talk was a breakthrough; it showed a clear path to establishing a link between electromagnetic duality and the Langlands Program. On the one hand, it brought into the realm of modern math a host of new ideas that mathematicians had not thought of (certainly not in connection with geometric Langlands): categories of branes, the special role played by the Hitchin moduli spaces in the Langlands Program, and the connection between A-branes and automorphic sheaves. On the other hand, this link also enabled physicists to use mathematical ideas and insights to advance the understanding of quantum physics.

Over the next two years, Witten worked out the details of his proposal, in collaboration with a Russian born physicist from Caltech, Anton Kapustin. Their paper[13] on this subject (230 pages long) appeared in April 2006 and made a big splash in both physics and mathematics communities. The opening paragraph from this paper describes many of the concepts that we have discussed in this book:

> The Langlands program for number fields unifies many classical and contemporary results in number theory and is a vast area of research. It has an analog for curves over a finite field, which has also been the subject of much celebrated work. In addition, a geometric version of the Langlands program for curves has been much developed, both for curves over a field of characteristic p and for ordinary complex Riemann surfaces.... Our focus in the present paper is on the geometric Langlands program for complex Riemann surfaces. We aim to show how this program can be understood as a chapter in quantum field theory. No prior familiarity with the Langlands program is assumed; instead, we assume a familiarity with subjects such as supersymmetric gauge theories, electric-magnetic duality, sigma models, mirror symmetry, branes, and topological field theory. The theme of the paper is to show that when these familiar physical ingredients are applied to just the right problem, the geometric Langlands program arises naturally.

Later in the introduction, Kapustin and Witten credited our meeting at the Institute for Advanced Study (in particular, the talk given there by my former student David Ben-Zvi) as the starting point for their research.

In the main body of the paper, Kapustin and Witten developed further the ideas that Witten formulated at our Princeton meeting. In particular, they elucidated the structure of the A-branes and B-branes arising in this picture, the mirror symmetry between them, and the link between the A-branes and the automorphic sheaves.

To explain their results, let us start with a simpler example of mirror symmetry. In the work by Kapustin and Witten, the mirror symmetry is between two Hitchin moduli spaces and the corresponding sigma models. But now let us replace one of these moduli spaces with a two-dimensional torus.

Such a torus may be viewed as the product of two circles. Indeed, the mesh on the picture shows clearly that the torus is like a bead necklace:

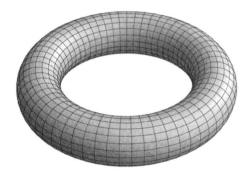

The role of the beads is played by the vertical circles in the mesh, and the role of the chainlet of the necklace, on which the beads are strung, is played by a horizontal circle, which we can imagine going through the middle of the torus. A mathematician would say that the necklace is a "fibration," whose "fibers" are beads and whose "base" is the chainlet. By the same token, the torus is a fibration whose fibers are circles and whose base is also a circle.

Let us call the radius of the base circle (chainlet) R_1 and the radius of the fiber circles (beads) R_2. It turns out that the mirror dual manifold will also be a torus. But it will be the product of the circles of radii $1/R_1$ and R_2. This inversion of the radius is similar to the inversion of the electric charge that happens under the electromagnetic duality.

So now we have two mirror dual tori – one of them, call it T, with the radii R_1 and R_2, and the other, call it $T^\vee$, with the radii $1/R_1$ and R_2. Note that if the base circle in T is big (that is, R_1 is large), then the base circle in $T^\vee$ is small (because then $1/R_1$ is small), and vice versa. This kind of switch between "big" and "small" is typical of all dualities in quantum physics.

Let's study B-branes on T and A-branes on $T^\vee$. They are matched under mirror symmetry, and this relation is well-understood (sometimes it is referred to as "T–duality" – T for torus).[14]

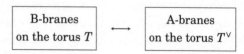

A typical example of a B-brane on the torus T is a so-called zero-brane, which is concentrated at a point p of T. It turns out that the dual A-brane on $T^\vee$, in contrast, will be "smeared" all over the torus $T^\vee$. What we mean by "smeared" requires an explanation. Without getting into too much detail, which would take us too far afield, this A-brane on $T^\vee$ is the torus $T^\vee$ itself equipped with an additional structure: a representation of its fundamental group in the circle group (similar to those we discussed in Chapter 15). This representation is determined by the position of the original point p in the torus T, so that, in fact, there is a one-to-one correspondence between the zero-branes on T and the A-branes "smeared" on $T^\vee$.

This phenomenon is similar to what happens under the so-called Fourier transform widely used in signal processing. If we apply Fourier transform to a signal concentrated near a particular moment in time, we obtain a signal that looks like a wave. The latter is "smeared" over the line that represents time, as shown on the picture.

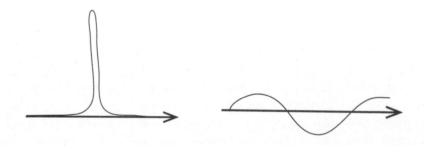

The Fourier transform may also be applied to many other types of signals, and there is an inverse transform, which allows us to recover the original signal. Often, complicated signals are transformed into simple ones, and that's why Fourier transform is useful in applications. Likewise, under mirror symmetry, complicated branes on one torus correspond to simple ones on the other, and vice versa.

It turns out that we can use this toric mirror symmetry to describe the mirror symmetry between the branes on the two Hitchin moduli spaces. Here we need to use an important property of these moduli spaces, which was described by Hitchin himself. Namely, the Hitchin moduli space is a fibration. The base of the fibration is a vector space, and the fibers are tori. That is, the whole space is a collection of tori, one for each point of the base. In the simplest case, both the base and the toric fibers are two-dimensional, and the fibration looks like this (note that the fibers at different points of the base may have different sizes):

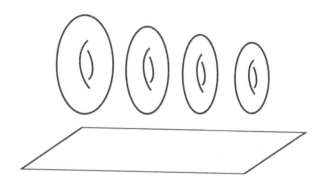

Think of the Hitchin fibration as a box of donuts, except that there are donuts attached not only to a grid of points in the base of the carton box, but to *all* points in the base. So we have infinitely many donuts – Homer Simpson would sure love that!

It turns out that the mirror dual Hitchin moduli space, the one associated to the Langlands dual group, is also a donut/toric fibration over the same base. ("Donuts. Is there anything they can't do?") This means that over each point in this base, we have two toric fibers: one in the Hitchin moduli space on the A-model side, and one in the Hitchin moduli space on the B-model side. Moreover, these two tori are mirror dual to each other, in the sense described above (if one of them has radii R_1 and R_2, then the other one has radii $1/R_1$ and R_2).

This observation gives us the opportunity to study mirror symmetry between two dual Hitchin moduli spaces fiber-wise, using the mirror symmetry between the dual toric fibers.

For example, let p be a point of the Hitchin moduli space $M(X, {}^L G)$. Let's take the zero-brane concentrated at this point. What will be the mirror dual A-brane on $M(X, G)$?

The point p belongs to a torus, which is the fiber of $M(X, {}^L G)$ over a point b in the base (the left torus on the picture below, on the B-model side). Consider the dual torus, which is the fiber of $M(X, G)$ over the same point b (the right torus on the picture, on the A-model side). The dual A-brane on $M(X, G)$ we are looking for will be the A-brane "smeared" over this dual torus. It will be the same dual brane as the one we obtain under the mirror symmetry between these two tori.

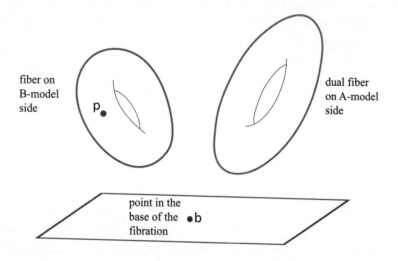

This kind of fiber-wise description of mirror symmetry – using dual toric fibrations – had been suggested earlier by Andrew Strominger, Shing-Tung Yau, and Eric Zaslow in a more general situation. It is now referred to as the SYZ conjecture, or SYZ mechanism.[15] This is a powerful idea: while mirror symmetry for dual tori is very well understood, mirror symmetry for general manifolds (such as the Hitchin moduli spaces) still appears quite mysterious. Therefore we can get a lot of mileage by reducing it to the toric case. Of course, to be able to implement it, we need to represent two mirror dual manifolds as dual toric fibrations over the same base (these fibrations also have to satisfy certain conditions). Luckily, we do have such fibrations in the case of the Hitchin moduli spaces, so we can put the SYZ mechanism to work. (In general, the dimensions of the toric fibers are greater than two, but the picture is similar.[16])

Now we use this mirror symmetry to construct the Langlands relation. First, it turns out that points of the Hitchin moduli space $M(X, {}^L G)$ are precisely the representations of the fundamental group of the Riemann surface X in ${}^L G$ (see endnote 1 to this chapter). Let's take the zero-brane concentrated at this point. According to the SYZ mechanism, the dual A-brane will be "smeared" over the dual torus (the fiber in the dual Hitchin moduli space over the same point in the base).

Kapustin and Witten not only described these A-branes in detail, they also explained how to convert them into automorphic sheaves of the geometric Langlands relation. Therefore the Langlands relation is achieved via this flow chart:

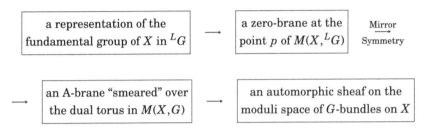

An essential element of this construction is the appearance of intermediate objects: A-branes. Kapustin and Witten proposed that the Langlands relation can be constructed in two steps: they first construct an A-brane using mirror symmetry. And then they construct an automorphic sheaf from this A-brane.[17] Up to now, we have only discussed the first step, the mirror symmetry. But the second step is also very interesting. In fact, the link between A-branes and automorphic sheaves was a groundbreaking insight of Kapustin and Witten; before their work, it was not known that there was such a link. Moreover, Kapustin and Witten suggested that a similar link exists in a far more general situation. This striking idea has already spurred a lot of mathematical research.

All this stuff, as my dad put it, is quite heavy: we've got Hitchin moduli spaces, mirror symmetry, A-branes, B-branes, automorphic sheaves... One can get a headache just trying to keep track of all of them. Believe me, even among specialists, very few people know the nuts and bolts of all elements of this construction. But my point is not for you to learn them all. Rather, I want to indicate the logical connections between these objects and show the creative process of scientists studying them: what drives them, how they learn from each other, how the knowledge they acquire is used to advance our understanding of the key questions.

But to lighten our loads somewhat, here is a diagram that illustrates the analogies between the objects we have discussed across the columns of Weil's Rosetta stone, plus an extra column corresponding to quantum physics. It extends the diagram on p. 161. (I have combined the left and the middle columns of Weil's Rosetta stone because the objects appearing in them are quite similar to each other.)

Number Theory & Curves/finite fields	Riemann surface X	Quantum Physics
Langlands relation	geometric Langlands relation	electromagnetic duality mirror symmetry
Galois group	fundamental group of X	fundamental group of X
representation of the Galois group in ^LG	representation of the fund. group in ^LG	zero-brane on $M(^LG,X)$
automorphic function	automorphic sheaf	A-brane on $M(G,X)$

Having looked at this diagram, my father asked me: How did Kapustin and Witten advance the Langlands Program? This is of course an important question. First of all, linking the Langlands Program to mirror symmetry and electromagnetic duality allows us to use the powerful arsenal of these areas of quantum physics to make new advances in the Langlands Program. Conversely, the ideas of the Langlands Program, transplanted into physics, motivated physicists to ask some questions about the electromagnetic duality that they had never asked before. This has already led to some fascinating discoveries. Second, the language of A-branes turns out to be well-adapted for the Langlands Program. Many of these A-branes have a much simpler structure than the automorphic sheaves, which are notoriously complicated. Therefore, using the language of A-branes, we can unveil some of the mysteries of the Langlands Program.

I want to show you a concrete example of how this new language can be applied. So let me tell you about my subsequent work[18] with Witten, which we completed in 2007. To explain what we did, I have to tell you about a problem that up to now I have sort of swept under the rug. In the above discussion, I pretended that all the fibers appearing in the two Hitchin moduli spaces are smooth tori that we are used to (like the ones shown on the above pictures – perfectly shaped donuts, if you will). In fact, this is only true for most of the fibers. But there are some special fibers that look different: they are degenerations of the smooth tori. If there were no degenerations, the SYZ mechanism would give us a complete description of the mirror symmetry between the branes on the two Hitchin moduli spaces. But the presence of the degenerate tori dramatically complicates this mirror symmetry. The most interesting and complicated part of the mirror symmetry is, in fact, what happens with the branes "living" on these degenerate tori.

Kapustin and Witten only considered in their paper the mirror symmetry restricted to the smooth tori. This left the question of the degenerate tori open. In our paper, Witten and I explained what happens in the case of the simplest degenerate tori, those with the so-called "orbifold singularities," such as this pinched torus:

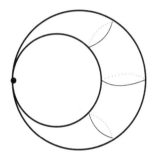

This is in fact the picture of a degenerate fiber arising in the case that our Riemann surface X is itself a torus, and the group ^{L}G is $SO(3)$ (it is taken directly from my paper with Witten). The base of the Hitchin fibration is a plane in this case. For every point on this plane, except for three special points, the fibers are the usual smooth tori. So outside these three points the Hitchin fibration is just a family of smooth tori/donuts. But in the neighborhood of each of the three special points, the "neck" of the toric fiber/donut collapses, as shown on the next picture, on which we keep track of the fibers over points within a given path in the base.

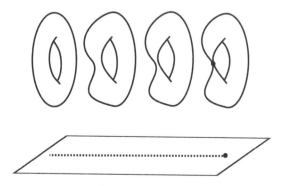

It's as if Homer Simpson got so excited about having a box with infinitely many donuts that he accidentally stepped on it, crushing some of the donuts (but don't worry about Homer; there would still be infinitely many perfectly shaped donuts left).

As we approach the marked point in the base (which is one of the three special points in the base), the neck of the torus in the fiber becomes thinner and thinner, until it collapses at the marked point. The fiber at the marked point is shown under a different angle on the picture above. It is not a torus anymore; it is what we could call a "degenerate" torus.

The question we need to answer is what happens when the zero-brane on the Hitchin moduli space is concentrated at the special point of the degenerate torus like the one marked on the above picture, at which the neck collapses. Mathematicians call it an orbifold singularity.

It turns out that this point has an additional symmetry group. In the example shown above, it is the same as the group of symmetries of a butterfly. In other words, it consists of the identity element and another element, corresponding to flipping the wings of the butterfly. This implies that there is not one, but two different zero-branes concentrated at this point. The question then is: what will be the corresponding two A-branes on the mirror dual Hitchin moduli space? (Note that in this case, G will be the group $SU(2)$, which is the Langlands dual group of $SO(3)$.)

As Witten and I explain in our paper, at each of the three special points in the base of the Hitchin fibration, the degenerate torus on the mirror dual side will look like this (the picture is taken from our paper):

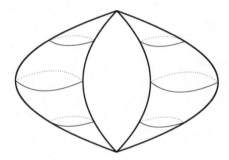

It appears in the Hitchin fibration in a similar fashion to what was shown on the previous picture, except that now, as we approach one of the special points in the base, the neck of the torus in the fiber becomes thinner and thinner at two places, and the neck collapses at both of them when we reach the special point on the base.

The corresponding degenerate fiber is quite different from the previous one because now the torus collapses at two points instead of one. Hence this degenerate torus has two pieces. Mathematicians call them components. Now we can answer our question: the two A-branes we are looking for (mirror dual to the two zero-branes concentrated at the singular point of the first degenerate torus) will be the A-branes "smeared" on each of the two components of the dual degenerate torus.

This is a prototype for what happens in general. When we look at the two Hitchin moduli spaces as fibrations over the same base, there will be degenerate fibers on both sides. But the mechanisms of degeneration will be different: if on the B-model side there is an orbifold singularity with an inner symmetry group (like the butterfly group in the above example), then the fiber on the A-model side will consist of several components, like the two components on the above picture. It turns out that there will be as many of these components as the number of elements in the symmetry group on the B-model side. This ensures that the zero-branes concentrated at the singular points precisely match the A-branes "smeared" on those different components.

In my paper with Witten, we analyzed this phenomenon in detail. Somewhat surprisingly, this led us to some new insights not only into the geometric Langlands Program for Riemann surfaces, but also into the middle column of Weil's Rosetta stone, which is about curves over finite fields. This is a good example of how ideas and insights in one area (quantum physics) could propagate back all the way to the roots of the Langlands Program.

Hereby lies the power of these connections. We now have not three, but four columns in Weil's Rosetta stone: the fourth one is for quantum physics. When we discover something new in this fourth column, we look at what the analogous results should be in the other three columns, and this may well become a source of new ideas and insights.

Witten and I started working on this project in April 2007 when I was visiting the Institute in Princeton, and the paper was finished on Halloween, October 31. (I remember the date well, because after posting it online I went to a Halloween party to celebrate.) During these seven months, I came to the Institute three times, each time for about a week. Every day we would work together at Witten's comfortable office. The rest of the time we were at different locations. I split my time between Berkeley and Paris in those days, and I also spent a couple of weeks visiting a math institute in Rio de Janeiro. But my where-abouts didn't matter. As long as I had a working Internet connection, we could collaborate effectively. During the most intense periods, we would exchange a dozen e-mails a day, ponder questions, send each other drafts of the paper, etc. Since we share the same first name, there was a kind of mirror symmetry be-tween our e-mails: each would start with "Dear Edward" and end with "Best, Edward."

This collaboration gave me the opportunity to observe Witten up close. I was amazed by both his intellectual power and work ethics. I sensed that he gives a lot of thought to the choice of a problem to work on. I have talked about this earlier in the book: there are problems that may take 350 years to solve, so it is important to estimate the ratio of importance of a given problem to the probability of success within a reasonable period of time. I think Witten has a great intuition for this, as well as great taste. And once he chooses the problem, he is relentless in pursuing it, like Tom Cruise's character in the film *Collateral*. His approach is thorough, methodical, with no stone left unturned. Like everyone else, he gets perplexed and confused from time to time. But he always finds his way. Working with him was inspiring and enriching on many levels.

The research on the interface between the Langlands Program and electro-magnetic duality quickly became a hot topic that developed into a vibrant area of research. An important role in this process was played by the annual con-ferences we organized at the Kavli Institue for Theoretical Physics in Santa Barbara. The Institute director, Nobel Prize–winner David Gross, was our big supporter.

In June 2009, I was asked to speak about these new developments at *Sémi-naire Bourbaki*. One of the longest running mathematics seminars in the world, it is revered in the mathematical community. Scores of mathematicians are drawn to its meetings at the Henri Poincaré Institute in Paris, which take up a weekend three times a year. The seminar was created shortly after World War II by a group of young and ambitious mathematicians who called them-selves – using an assumed name – *Association des collaborateurs de Nicolas Bourbaki*. Their idea was to rewrite the foundations of mathematics using a new standard of rigor based on the set theory initiated by Georg Cantor in the late nineteenth century. They succeeded only partially, but their influence on mathematics has been enormous. André Weil was one of the founding mem-bers, and Alexander Grothendieck played a big role later on.

The purpose of *Séminaire Bourbaki* is to report on the most exciting developments in mathematics. A secret committee that chooses the topics and the speakers has since its inception followed the rule that its members must be under fifty years of age. The founders of the Bourbaki movement apparently believed that it constantly needed fresh blood, and this has served them well. The committee invites the speakers and makes sure that they write up their lectures in advance. Copies are distributed to the audience at the seminar. As it is considered an honor to present a talk at the seminar, the speakers comply with the request.

The title of my seminar was "Gauge Theory and Langlands Duality."[19] Though my talk was more technical and involved more formulas and mathematical terminology, I basically followed the story that I have recounted in this book. I started with André Weil's Rosetta stone, giving a brief tour of each of its three columns, just as I did here. Because Weil was one of the founders of the Bourbaki group, I thought it was especially fitting to talk about his ideas at the seminar. I then focused on the newest developments, linking the Langlands Program and the electromagnetic duality.

The talk was received well. I was pleased to see in the front row another key member of the Bourbaki, Jean-Pierre Serre, a legend in his own right. At the end of my talk he came up to me. After asking a few pointed technical questions, he made an observation.

"I found it interesting that you think of quantum physics as the fourth column in Weil's Rosetta stone," he said. "You know, André Weil wasn't particularly fond of physics. But I think that if he were here today, he would agree that quantum physics has an important role to play in this story."

This was the best compliment one could possibly get.

In the last few years, a lot of progress has been made in the Langlands Program, across all columns of Weil's Rosetta stone. We are still far from fully understanding the deepest mysteries of the Langlands Program, but one thing is clear: it has passed the test of time. We see more clearly now that it has led us to some of the most fundamental questions in math and physics.

These ideas are as vital today as they were when Langlands wrote his letter to André Weil, almost fifty years ago. I don't know whether we can find all the answers in the next fifty years, but there is no doubt that the next fifty years will be at least as exciting as the past fifty years have been. And perhaps some of the readers of this book will have the opportunity to contribute to this fascinating project.

The Langlands Program has been the focus of this book. I think it provides a good panoramic view of modern mathematics: its deep conceptual structure, groundbreaking insights, tantalizing conjectures, profound theorems, and unexpected connections between different fields. It also illustrates the intricate links between math and physics and the mutually enriching dialogue between these two subjects. Thus, the Langlands Program exemplifies the four qualities of mathematical theories that we discussed in Chapter 2: universality, objectivity, endurance, and relevance to the physical world.

Of course, there are many other fascinating areas of math. Some have been exposed in the literature for non-specialists and some have not. As Henry David Thoreau wrote,[20] "We have heard about the poetry of mathematics, but very little of it has yet been sung." Alas, his words still ring true today, more than 150 years after he wrote them, which is to say that we, mathematicians, need to do a better job of unlocking the power and beauty of our subject to a wider audience. At the same time, I hope that the story of the Langlands Program will inspire readers' curiosity about mathematics and motivate the desire to learn more.

Chapter 18

Searching for the Formula of Love

In 2008, I was invited to do research and lecture about my work in Paris as the recipient of a newly created *Chaire d'Excellence* awarded by Fondation Sciences Mathématiques de Paris.

Paris is one of the world's centers of mathematics; it is also a capital of cinema. Being there, I felt inspired to make a movie about math. In popular films, mathematicians are usually portrayed as weirdos and social misfits on the verge of mental illness, reinforcing the stereotype of mathematics as a boring and cold subject, far removed from reality. Who could want such a life for themselves, doing work that supposedly had nothing to do with anything?

When I came back to Berkeley in December 2008, I felt the urge to channel my artistic energy. My neighbor, Thomas Farber, is a wonderful author, and he teaches creative writing at UC Berkeley. I asked him, "How about writing a screenplay together about a writer and a mathematician?" Tom liked the idea and suggested that we set it on a beach in the south of France. We decided that the film would start like this: a writer and a mathematician, on a beautiful sunny day, sitting at adjacent tables in an open-air café on the beach. They savor the beauty surrounding them, look at each other, and start talking. What happens next?

We began writing. The process was similar to the way I collaborate with mathematicians and physicists. But it was also different: finding the right words to describe the characters' feelings and emotions, getting to the heart of a story. The framework was much more fluid and unrestricted than I had been used to. And there I was, going toe-to-toe with a great writer for whom I had so much respect and admiration. Luckily for me, Tom did not try to impose his will but treated me as an equal, gently letting me develop my abilities as a writer. Like the mentors who guided me into the world of mathematics, Tom helped me to enter the world of writing, for which I will always be grateful.

In one of the conversations, the mathematician tells the writer about the "two-body problem." It refers to two objects (bodies) that interact only with each other, such as a star and a planet (we ignore all other forces acting on them). There is a simple mathematical formula that accurately predicts their trajectories all the way into the future once we know the force of attraction between them. How different, however, from the interaction of two human bodies – two lovers, or two friends. Here, even if the two-body problem has a solution, it's not unique.

Our screenplay is about the collision between the real world and the world of abstraction: for Richard, the writer, it's the world of literature and art; for Phillip, the mathematician, the world of science and mathematics. Each man is fluent in his respective abstract domain, but in what way does this affect behavior in the real world? Phillip is trying to come to terms with a dichotomy between mathematical truth, where he is expert, and human truth, where he is not. He learns that approaching life's problems in the same way as mathematical problems does not always help.

Tom and I also asked: can we see the differences and similarities between art and science – the "two cultures," as C.P. Snow called them[1] – through the narratives of these two men? In fact, one can read the film as metaphorical, about two sides of the same character: the left half of the brain and the right half, if you will. They are in constant competition but also inform each other – two cultures co-existing in one mind.

In our screenplay, the characters trade stories of their past relationships, love found and lost, heartbreaks. And they meet several women in the course of the day, so we can see the two men use their passion for their professions as a means of seduction. There is a lot of mutual interest between them as well, but at the same time a conflict is brewing, which reaches an unexpected conclusion at the end.

We called our screenplay *The Two-Body Problem* and published it as a book.[2] Its play version has been performed in a Berkeley theater, directed by award-winning director Barbara Oliver. This was the first time I had ventured into the arts, and I was both surprised and amused by the audience reaction. For example, most people took everything that happened to the mathematician in the screenplay as my autobiography. Of course, many of my real-life experiences contributed to the writing of *The Two-Body Problem*. For instance, I did have a Russian girlfriend in Paris, and some of the remarkable qualities of Natalia, Phillip's girlfriend in the screenplay, were inspired by her. Some scenes in the screenplay were drawn from my experience, some from Tom's. But as a writer, you are most driven by the desire to create compelling characters and an engaging story. Once Tom and I decided what we wanted to convey, we had to shape the characters in a certain way. Those real-life experiences got so embellished and distorted that they were no longer ours. The protagonists of *The Two-Body Problem* became their own men, as they had to, to be art.

As we began looking for a producer to help us make *The Two-Body Problem* into a full-length feature film, I thought it would be worthwhile to do a cinematic project on a smaller scale. When I returned to Paris to continue my *Chaire d'Excellence* in April 2009, a friend, mathematician Pierre Schapira, introduced me to a young, talented film director, Reine Graves. A former fashion model, she had previously directed several original, bold short films (one of which won the Pasolini Prize at the Festival of Censored Films in Paris). At a lunch meeting arranged by Pierre, she and I hit it off right away. I suggested we work together on a short film about math, and Reine liked the idea. Months later, when asked about this, she said that she felt mathematics was one of the last remaining areas where there was genuine passion.[3]

As we started throwing ideas around, I showed Reine a couple of photographs I had made previously, in which I painted (digitally) tattoos of mathematical formulas on human bodies. Reine liked them, and we decided that we would try to make a film involving the tattoo of a formula.

Tattoo, as an art form, originated in Japan. I have visited Japan a dozen times (to work with Feigin, who had been spending his summers at Kyoto University) and am fascinated by Japanese culture. Not surprisingly, Reine and I turned to the Japanese cinema for inspiration. One film was *Rite of Love and Death* by the great Japanese writer Yukio Mishima, based on his short story. Mishima himself directed and starred in it.

The film is shot in black-and-white, and it unfolds on the austere, stylized stage typical of Japanese Noh theater. The film has no dialogue, but there is music from Wagner's opera *Tristan and Isolde* playing in the background. There are two characters: a young officer of the Imperial Guard, Lieutenant Takeyama, and his wife, Reiko. The officer's friends stage an unsuccessful *coup d'état* (here the film refers to actual events of February 1936, which Mishima thought had a dramatic effect on Japanese history). The lieutenant is given the order to execute the perpetrators of the *coup*, which he cannot do – they are close friends. But neither can he disobey the order of the Emperor. The only way out is ritual suicide, *seppuku* (or *harakiri*).

Although it is only twenty-nine minutes long, the film touched me deeply. I could sense the vigor and clarity of Mishima's vision. His presentation was forceful, raw, unapologetic. One may disagree with his ideas (and in fact his vision of the intimate link between love and death does not appeal to me), but I have a tremendous respect for the author for being so strong and uncompromising.

Mishima's film went against the usual conventions of cinema: it was silent, with written text between the "chapters" of the movie to explain what's going to happen next. It was theatrical; scenes carefully staged, with little movement. But I was captivated by the undercurrent of emotion. (I did not know yet the eerie resemblance of Mishima's own death to what happened in his film.)

Perhaps the film resonated with me so much because Reine and I were also trying to create an unconventional film, to talk about mathematics the way no one had talked about it before. I felt that Mishima had created the aesthetic framework and language we were looking for. I called Reine.

"I have watched Mishima's film," I said, "and it's amazing. We should make a film like this."

"OK," she said, "but what will it be about?"

Suddenly, words started coming out of my mouth. Everything was crystal clear.

"A mathematician creates a formula of love," I said, "but then discovers the flip side of the formula: it can be used for evil as well as for good. He realizes he has to hide the formula to protect it from falling into the wrong hands. And he decides to tattoo it on the body of the woman he loves."

"Sounds good. What do you think we should call it?"

"Hmmm... How about this: *Rites of Love and Math*."

And just like this, the idea of the film was born.

We envisioned it as an allegory, showing that a mathematical formula can be beautiful, like a poem, a painting, or a piece of music. The idea was to appeal not to the cerebral but rather to the intuitive and visceral. Let the viewers first *feel* rather than *understand* it. We thought emphasizing the human and spiritual elements of mathematics would help inspire viewer's curiosity.

Mathematics and science in general are often presented as cold and sterile.

In truth, the process of creating new mathematics is a passionate pursuit, a deeply personal experience, just like creating art and music. It requires love and dedication, a struggle with the unknown and with oneself, which elicits strong emotions. And the formulas you discover really do get under your skin, just like the tattooing in the film.

In our film, a mathematician discovers a "formula of love." Of course, this is a metaphor: we are always trying to achieve complete understanding, ultimate clarity, to know everything. In the real world, we have to settle for partial knowledge and understanding. But what if someone were able to find the ultimate Truth; what if it could be expressed by a mathematical formula? This would be the formula of love.

Henry David Thoreau put this eloquently:[4]

> The most distinct and beautiful statement of any truth must take at last the mathematical form. We might so simplify the rules of moral philosophy, as well as of arithmetic, that one formula would express them both.

Even if a single formula can't be powerful enough to explain everything, mathematical formulas are some of the purest, most versatile, and most economical expressions of truth known to mankind. They convey timeless and precious knowledge, unaffected by fads and fashion, and impart the same meaning to anyone who comes in contact with them. The truths they express are the necessary truths, steadfast beacons of reality guiding humanity through time and space.

Heinrich Hertz, who proved the existence of electromagnetic waves and whose name is now used as the unit of frequency, expressed his awe this way:[5] "One cannot escape the feeling that these mathematical formulas have an independent existence and an intelligence of their own, that they are wiser than we are, wiser even than their discoverers."

Hertz is not alone in his sentiment. Most math practitioners believe that mathematical formulas and ideas inhabit a separate world. Robert Langlands writes that mathematics "often comes in the form of intimations, a word that suggests that mathematics, and not only its basic concepts, exists independently of us. This is a notion that is hard to credit, but hard for a professional mathematician to do without."[6] This is echoed by another eminent mathematician, Yuri Manin (the advisor of Drinfeld), who talks about a "vision of the great Castle of Mathematics, towering somewhere in the Platonic World of Ideas, which [mathematicians] humbly and devotedly discover (rather than invent)."[7]

From this point of view, Galois groups were *discovered* by the French prodigy, not *invented* by him. Until he did so, this concept lived somewhere in the enchanted gardens of the ideal world of mathematics, waiting to be found. Even if Galois' papers had been lost and he had not been given the rightfully deserved credit for his discovery, exactly the same groups would have been discovered by someone else.

Contrast this with discoveries in other areas of human endeavor: if Steve Jobs had not come back to Apple, we may have never known iPods, iPhones, and iPads. Other technological innovations would have been made, but there is no reason to expect that the same elements would have been found by others. In contrast, mathematical truths are inevitable.

The world inhabited by mathematical concepts and ideas is often referred to as the Platonic world of mathematics, after the Greek philosopher Plato, who was first to argue that mathematical entities are independent of our rational activities.[8] In his book *The Road to Reality*, acclaimed mathematical physicist Roger Penrose writes that the mathematical assertions that belong to the Platonic world of mathematics "are precisely those that are objectively true. To say that some mathematical assertion has a Platonic existence is merely to say that it is true in an objective sense." Similarly, mathematical notions "have a Platonic existence because they are objective notions."[9]

Like Penrose, I believe that the Platonic world of mathematics is separate from both the physical world and the mental world. For example, consider Fermat's Last Theorem. Penrose asks rhetorically in his book whether "we take the view that Fermat's assertion was always true, long before Fermat actually made it, or [that] its validity [is] a purely cultural matter, dependent upon whatever might be the subjective standards of the community of human mathematicians?"[10] Relying on the time-honored tradition of argument by *reductio ad absurdum*, Penrose shows that embracing the subjective interpretation quickly leads us to assertions that are "patently absurd," underscoring the independence of mathematical knowledge of any human activity.

Kurt Gödel, whose work – especially, the celebrated incompleteness theorems – revolutionized mathematical logic, was an unabashed proponent of this view. He wrote that mathematical concepts "form an objective reality of their own, which we cannot create or change, but only perceive and describe."[11] In other words, "mathematics describes a non-sensual reality, which exists independently both of the acts and of the dispositions of the human mind and is only perceived, and probably perceived very incompletely, by the human mind."[12]

The Platonic world of mathematics also exists independently of physical reality. For example, as we discussed in Chapter 16, the apparatus of gauge theories was originally developed by mathematicians without any reference to physics. In fact, it turns out that only three of those models describe known forces of nature (electromagnetic, weak, and strong). They correspond to three specific Lie groups (the circle group, $SU(2)$, and $SU(3)$, respectively), even though there is a gauge theory for *any* Lie group. The gauge theories associated to the Lie groups other than those three are perfectly sound mathematically, but there are no known connections between them and the real world. Furthermore, we have talked about the supersymmetric extensions of these gauge theories, which we can analyze mathematically even though supersymmetry has not been found in nature, and quite possibly is not present there at all. Similar models also make sense mathematically in a space-time that has dimension different from four. There are plenty of other examples of rich mathematical theories that are not directly linked to any kind of physical reality.

In his book *Shadows of the Mind*, Roger Penrose talks about the triangle: the physical world, the mental world, and the Platonic world of math.[13] They are separate but deeply intertwined with each other. We still don't fully understand how they are linked together, but one thing is clear: each of them affects our lives in profound ways. However, while we all appreciate the significance of the physical and mental worlds, many of us remain blissfully ignorant of the world of mathematics. I believe that when we awaken to this hidden reality and use its untapped powers, this will lead to a shift in our society on the order of the Industrial Revolution.

In my view, it is the objectivity of mathematical knowledge that is the source of its limitless possibilities. This quality distinguishes mathematics from any other type of human endeavor. I believe that understanding what is behind this quality will shed light on the deepest mysteries of physical reality, consciousness, and interrelations between them. In other words, the closer we are to the Platonic world of math, the more power we will have to understand the world around us and our place in it.

Luckily, nothing can stop us from delving deeper into this Platonic reality and integrating it into our lives. What's truly remarkable is mathematics' inherent democracy: while some parts of the physical and mental worlds may be perceived or interpreted differently by different people or may not even be accessible to some of us, mathematical concepts and equations are perceived in the same way and belong to all of us *equally*. No one can have a monopoly on mathematical knowledge; no one can claim a mathematical formula or idea as his or her invention; no one can patent a formula! Albert Einstein, for example, would not be able to patent his formula $E = mc^2$. That's because, if correct, a mathematical formula expresses an eternal truth about the universe. Hence no one can claim ownership of it; it is ours to share.[14] Rich or poor, black or white, young or old – no one can take these formulas away from us. Nothing in this world is so profound and elegant, and yet so available to all.

Following Mishima, the centerpiece of the austere decor in *Rites of Love and Math* was a large calligraphy hanging on the wall. In Mishima's film the calligraphy read *shisei*: sincerity. His film was about sincerity and honor. Ours was about the truth, so naturally we thought our calligraphy should say truth. And we decided to do it not in Japanese, but in Russian.

The word "truth" has two translations into Russian. The more familiar *pravda* refers to factual truth, such as a news item (hence the name of the official newspaper of the Communist Party of the USSR). The other one, *istina*, means deeper, philosophical truth. For example, the statement that the group of symmetries of a round table is a circle is *pravda*, but the statement of the Langlands Program (in the cases in which it has been proved) is *istina*. Clearly, the truth for which the Mathematician sacrifices himself in the movie is *istina*.

In our film, we wanted to reflect on the moral aspect of mathematical knowledge: a formula with so much power may well have a flip side and could potentially be used for evil. Think of a group of theoretical physicists at the beginning of the twentieth century trying to understand the structure of the atom. What they thought was a pure and noble scientific pursuit inadvertently led them to the discovery of atomic energy. It brought us a lot of good, but also destruction and death. Likewise, a mathematical formula discovered as part of our quest for knowledge could prove to be harmful. Although scientists should be free to pursue their ideas, I also believe that it is our responsibility to do everything in our power to ensure that the formulas we discover are not used for evil. That's why in our film the Mathematician is prepared to die to protect the formula from falling into the wrong hands. Tattooing is his way to hide the formula and at the same time ensure that it survives.

Because I have never had a tattoo, I had to learn about the process. These days tattoos are made with a machine, but historically (in Japan) tattoos were engraved with a bamboo stick – a longer, more painful process. I've been told it's still possible to find tattoo parlors in Japan that use this ancient technique. This is how we presented it in the film.

Which formula should play the role of "formula of love" was a big question. It had to be sufficiently complicated (it's a formula of love, after all) but also aesthetically pleasing. We wanted to convey that a mathematical formula could be beautiful in content as well as form. And I wanted it to be *my* formula.

Doing "casting" for the formula of love, I stumbled on this:

$$\int_{\mathbb{CP}^1} \omega F(qz,\overline{qz}) = \sum_{m,\overline{m}=0}^{\infty} \int_{|z|<\epsilon^{-1}} \omega_{z\overline{z}}\, z^m \overline{z}^{\overline{m}} dz d\overline{z} \cdot \frac{q^m \overline{q}^{\overline{m}}}{m!\overline{m}!} \partial_z^m \partial_{\overline{z}}^{\overline{m}} F \bigg|_{z=0}$$

$$+ q\overline{q} \sum_{m,\overline{m}=0}^{\infty} \frac{q^m \overline{q}^{\overline{m}}}{m!\overline{m}!} \partial_w^m \partial_{\overline{w}}^{\overline{m}} \omega_{w\overline{w}} \bigg|_{w=0} \cdot \int_{|w|<q^{-1}\epsilon^{-1}} F\, w^m \overline{w}^{\overline{m}} dw d\overline{w}.$$

It appears as formula (5.7) in a hundred-page paper *Instantons Beyond Topological Theory I*, which I wrote in 2006 with two good friends, Andrey Losev and Nikita Nekrasov.[15]

This equation can seem forbidding enough that if we had made a film in which I wrote this formula on a blackboard and tried to explain its meaning, most people would have probably walked out of the theater. But seeing it in the form of a tattoo elicited a totally different reaction. It really got under everyone's skin: everyone wanted to know what it meant.

So what *does* it mean? Our paper was the first installment in a series we wrote about a new approach to quantum field theories with "instantons" – these are configurations of fields with remarkable properties. Although quantum field theories have been successful in accurately describing the interaction between elementary particles, there are many important phenomena that are still poorly understood. For example, according to the Standard Model, protons and neutrons consist of three quarks each, which cannot be separated. In physics, this phenomenon is known as confinement. Its proper theoretical explanation is still lacking, and many physicists believe that instantons hold the key to solving this mystery. However, in the conventional approach to quantum field theories, instantons are elusive.

We proposed a novel approach to quantum field theories that we hoped would help us understand better instantons' powerful effects. The above formula expresses a surprising identity between two ways to compute a correlation function in one of our theories.[16] Little did we know at the time we discovered it that it would soon be slated to play the role of formula of love.

Oriane Giraud, our special effects artist, liked the formula, but said it was too involved for a tattoo. I simplified the notation, and here's how it appears in our film:

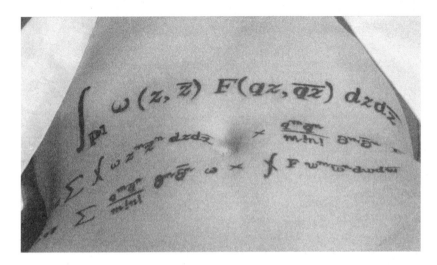

The tattoo scene in the film was meant to represent the passion involved in doing mathematical research. While he is making the tattoo, the Mathematician completely shuts himself off from the world. To him, the formula really becomes a question of life and death.

Shooting this scene took us many hours. It was psychologically and physically draining both for me and for Kayshonne Insixieng May, the actress playing Mariko. We finished this scene close to midnight on our last day of shooting. It was an emotional moment for our crew of about thirty, after everything we had been through together.

The film's premiere was in April 2010, sponsored by Fondation Sciences Mathématiques de Paris, at Max Linder Panorama theater, one of best in Paris. It was a success. The first articles about the film started to appear. *Le Monde* called *Rites of Love and Math* "a stunning short film" that "offers an unusual romantic vision of mathematicians."[17] And the *New Scientist* wrote:[18]

> It is beautiful to look at.... If Frenkel's goal was to bring more people to maths, he can congratulate himself on a job well done. The formula of love, which is actually a simplified version of an equation he published in a 2006 paper on quantum field theory entitled "Instantons beyond topological theory I," will probably soon have been seen – if not understood – by a far larger audience than it would otherwise ever have reached.

In the words of the popular French magazine *Tangente Sup*,[19] the film "will intrigue those who think of mathematics as the absolute opposite of art and poetry." In an insert accompanying the article, Hervé Lehning wrote:

> In the mathematical research of Edward Frenkel, symmetry and duality are of great importance. They are related to the Langlands Program which aims to establish a bridge between the theory of numbers and representations of certain groups. This very abstract subject actually has applications, for example in cryptography.... If the idea of duality is so important to Edward Frenkel, one could ask whether he sees a duality between love and mathematics, as the title of his film would suggest. His answer to this question is clear. For him, mathematical research is like a love story.

Since then, the film has been shown at film festivals in France, Spain, and California; in Paris, Kyoto, Madrid, Santa Barbara, Bilbao, Venice... The screenings and the ensuing publicity gave me the opportunity to see some of the differences between the "two cultures." At first, this came as culture shock. My mathematics can be fully understood only by a small number of people; sometimes, no more than a dozen in the whole world at first. Furthermore, because each mathematical formula represents an objective truth, there is in essence only one way to interpret that truth. My mathematical work is therefore perceived in the same way by everyone who reads it. In contrast, our film was intended for a wide audience: thousands were being exposed to it. And, of course, they all interpreted it in their own ways.

What I learned from this is that the viewer is always part of an artistic project; at the end of the day, it's all in the eye of the beholder. A creator has no power over viewers' perceptions. But of course, this is something we can benefit from because when we share our views we all get enriched.

In our film, we attempted to create a synthesis of the two cultures by speaking about mathematics with an artist's sensibility. At the beginning of the film, Mariko is writing a love poem to the Mathematician.[20] When, at the end of the film, he tattoos the formula, this is his way to reciprocate: for him, the formula is an expression of his love. It can carry the same passion and emotional charge as a poem, so this was our way to show the parallel between mathematics and poetry. For the Mathematician, it's his gift of love, the product of his creation, passion, imagination. It's as if he is writing a love letter to her – remember the young Galois writing his equations on the eve of his death.

But who is she? In the framework of the mythical world we envisioned, she is the incarnation of Mathematical Truth (hence her name Mariko, "truth" in Japanese, and that's why the word *istina* is calligraphed on the painting hanging on the wall). The Mathematician's love for her is meant to represent his love for Mathematics and Truth, for which he sacrifices himself. But she has to survive and carry his formula, as she would their child. Mathematical Truth is eternal.

Can mathematics be a language of love? Some viewers were uneasy about the idea of a "formula of love." For example, someone said to me after watching the film: "Logic and feelings don't always get along. That's why we say that love is blind. So how could a formula of love possibly work?" Indeed, our feelings and emotions often appear to us as irrational (though cognitive scientists will tell you that some aspects of this apparent irrationality can actually be described using mathematics). Therefore, I don't believe that there is a formula that describes or explains love. When I talk about a connection between love and math, I don't mean to say that love can be reduced to math. Rather, my point is that there is a lot more to math than most of us realize. Among other things, mathematics gives us a rationale and an additional capacity to love each other and the world around us. A mathematical formula does not explain love, but it can carry a charge of love.

As poet Norma Farber wrote,[21]

> Make me no lazy love...
> Move me from case to case.

Mathematics moves us "from case to case," and herein lies its deep and largely untapped spiritual function.

Albert Einstein wrote:[22] "Every one who is seriously involved in the pursuit of science becomes convinced that some spirit is manifest in the laws of the Universe – a spirit vastly superior to that of man, and one in the face of which we with our modest powers must feel humble." And Isaac Newton expressed his feelings this way:[23] "to myself I seem to have been only like a boy playing on the seashore, and diverting myself in now and then finding a smoother pebble or a prettier shell than ordinary, whilst the great ocean of truth lay all undiscovered before me."

My dream is that one day we will all awaken to this hidden reality. We may then perhaps be able to set aside our differences and focus on the profound truths that unite us. Then, we will all be like children playing on the seashore, marveling at the dazzling beauty and harmony we discover, share, and cherish together.

Epilogue

My plane is landing at Logan Airport in Boston. It is January 2012. I am coming to the Annual Joint Meeting of the American Mathematical Society (AMS) and Mathematical Association of America, invited to deliver the 2012 AMS Colloquium Lectures. These lectures have been given annually since 1896. Looking at the list of past speakers and the subjects of their lectures is like revisiting the history of mathematics of the past century: John von Neumann, Shiing-Shen Chern, Michael Atiyah, Raoul Bott, Robert Langlands, Edward Witten, and many other great mathematicians. I feel honored and humbled to be part of this tradition.

Coming back to Boston brings up memories. My first landing at Logan was in September 1989, when I came to Harvard – to paraphrase the famous movie title, *From Russia with Math*. I was twenty-one then, not knowing yet what to expect, what was to come. Three months later, growing up fast in those turbulent times, I was back at Logan to see off my mentor Boris Feigin who was returning to Moscow, wondering when I would see him again. In fact, our mathematical collaboration and friendship continued and flourished.

My stay at Harvard turned out to be much longer than I expected: getting my Ph.D. the following year, being elected to the Harvard Society of Fellows, and at the end of my term there, being appointed Associate Professor at Harvard. Then, five years after my arrival in Boston, I anxiously waited at Logan for my parents and my sister's family to arrive, to join me and settle in America. They have been living in the Boston area since then, but I left in 1997, after the University of California at Berkeley made me an offer I couldn't refuse.

I still visit Boston regularly to see my family. In fact, my parents' place is just a few blocks from the Hynes Convention Center, where the Joint Mathematics Meeting convenes, so they will have a chance to see me in action for the first time. What a beautiful gift – to be able to share this experience with my family. "Welcome home!"

The Joint Meeting has more than 7,000 registered participants – most likely, the biggest math gathering ever. Many of them came to my lectures, held in a giant ballroom. My parents, my sister, and my niece are in the front row. The lectures are about my recent joint work with Robert Langlands and Ngô Bao Châu. It is the result of our three-year collaboration, our attempt to develop further the ideas of the Langlands Program.[1]

"What if we were to make a film about the Langlands Program?" I ask the audience. "Then, as any screenwriter would tell you, we would have to grapple with questions like these: What's at stake? Who are the characters? What's the story line? What are the conflicts? How are they resolved?"

People in the audience are smiling. I talk about André Weil and his Rosetta stone. We go on a journey through different continents of the world of mathematics, examining mysterious connections between them.

Every click of the remote brings up the next slide of my presentation beamed to four giant screens. Each describes a small step in our never-ending quest for knowledge. We are pondering eternal questions of truth and beauty. And the more we learn about mathematics – this magic hidden universe – the more we realize how little we know, how much more lies ahead. Our journey continues.

Acknowledgments

I thank DARPA and the National Science Foundation for supporting some of my research described in this book. The book was completed while I was a Miller Professor at the Miller Institute for Basic Research in Science at UC Berkeley.

I thank my editor T.J. Kelleher and project editor Melissa Veronesi at Basic Books for their expert guidance.

While working on the book, I have benefitted from fruitful discussions with Sara Bershtel, Robert Brazell, David Eisenbud, Marc Gerald, Masako King, Susan Rabiner, Sasha Raskin, Philibert Schogt, Margit Schwab, Eric Weinstein, and David Yezzi.

I thank Alex Freedland, Ben Glass, Claude Levesque, Kayvan Mashayekh, and Corinne Trang for reading parts of the book at various stages and offering helpful advice. I am grateful to Andrea Young for taking the photos of the "cup trick" used in Chapter 15.

Special thanks are due to Thomas Farber for numerous insights and expert advice, and to Marie Levek for reading the manuscript and asking probing questions that helped me improve the presentation in many places. My father, Vladimir Frenkel, read the book's many drafts and his feedback was invaluable.

My debts to my teachers, mentors, and others who helped me on my journey are, I hope, clear from the story I've told.

Above all, my gratitude is for my parents, Lidia and Vladimir Frenkel, whose love and support made possible all I have achieved. I dedicate this book to them.

Notes

Preface

1. Edward Frenkel, *Don't Let Economists and Politicians Hack Your Math*, Slate, February 8, 2013, http://slate.me/128ygaM

Chapter 1. A Mysterious Beast

1. Image credit: Physics World, http://www.hk-phy.org/index2.html

2. Images credit: Arpad Horvath.

Chapter 2. The Essence of Symmetry

1. In this discussion we use the expression "symmetry of an object" as the term for a particular transformation preserving an object, such as a rotation of a table. We do not say "symmetry of an object" to express the property of an object to be symmetrical.

2. If we use clockwise rotation, we get the same set of rotations: clockwise rotation by 90 degrees is the same as the counterclockwise rotation by 270 degrees, etc. As a matter of convention, mathematicians usually consider counterclockwise rotations, but this is just a matter of choice.

3. This may seem superfluous, but we are not just being pedantic here. We must include it, if we are to be consistent. We said that a symmetry is any transformation that preserves our object, and the identity is such a transformation.

To avoid confusion, I want to stress that in this discussion we only care about the end result of a given symmetry. What we do to the object in the process does not matter; only the final positions of all points of the object do. For example, if we rotate the table by 360 degrees, then every point of the table ends up in the same position as it was initially. That's why for us rotation by 360 degrees is the same symmetry as no rotation at all. For the same reason, rotation by 90 degrees counterclockwise is the same as rotation by 270 degrees clockwise. As another example, suppose that we slide the table on the floor ten feet in a certain direction and then slide it back ten feet, or that we move the table to another room and then bring it back. As long as it ends up in the same position and each of its points ends up in the same position as it was initially, this is considered the same symmetry as the identical symmetry.

4. There is an important property that the composition of symmetries satisfies called *associativity*: given three symmetries, S, S', and S'', taking their composition in two different orders,

$(S \circ S') \circ S''$ and $S \circ (S' \circ S'')$, gives the same result. This property is included in the formal definition of a group as an additional axiom. We do not mention it in the main body of the book because for the groups we consider, this property is obviously satisfied.

5. When we talked about the symmetries of a square table, we found it convenient to identify the four symmetries with the four corners of the table. However, such an identification depends on the choice of one of the corners – the one that represents the identity symmetry. Once this choice is made, we can indeed identify each symmetry with the corner to which the chosen corner is transformed by this symmetry. The drawback is that if we choose a different corner to represent the identity symmetry, we obtain a different identification. Hence it is better to make a distinction between the symmetries of the table and the points of the table.

6. See Sean Carroll, *The Particle at the End of the Universe: How the Hunt for the Higgs Boson Leads Us to the Edge of a New World*, Dutton, 2012.

7. Mathematician Felix Klein used the idea that shapes are determined by their symmetry properties as the point of departure for his highly influential Erlangen Program in 1872, in which he declared that salient features of any geometry are determined by a symmetry group. For example, in Euclidean geometry, the symmetry group consists of all transformations of the Euclidean space that preserve distances. These transformations are compositions of rotations and translations. Non-Euclidean geometries correspond to other symmetry groups. This allows us to classify possible geometries by classifying relevant symmetry groups.

8. This is not to say that no aspects of a mathematical statement are subject to interpretation; for instance, questions like how important a given statement is, how widely applicable, how consequential for the development of mathematics, and so on, could be subject to a debate. But the *meaning* of the statement – what exactly it says – is not open to interpretation if the statement is logically consistent. (The logical consistency of the statement is not subject to a debate, either, once we choose the system of axioms within which the statement is made.)

9. Note that each rotation also gives rise to a symmetry of any round object, such as a round table. Therefore, in principle, one could also speak of a representation of the group of rotations by symmetries of the round table rather than a plane. However, in mathematics the term "representation" is reserved specifically for the situation in which a given group gives rise to symmetries of an n-dimensional space. These symmetries are required to be what mathematicians call linear transformations, a concept explained in endnote 2 to Chapter 14.

10. For any element g of the group of rotations, denote the corresponding symmetry of the n-dimensional space by S_g. It has to be a linear transformation for any g, and the following properties must be satisfied: first, for any pair of elements of the group, g and h, the symmetry $S_{g \cdot h}$ must be equal to the composition of the symmetries S_g and S_h. And second, the symmetry corresponding to the identity element of the group must be the identity symmetry of the plane.

11. Later on, it was discovered that there are three more quarks, called "charm," "top," and "bottom," and the corresponding anti-quarks.

Chapter 3. The Fifth Problem

1. There was also a small semi-official synagogue in Marina Rosha. The situation improved after *perestroika* as more synagogues and Jewish community centers opened in Moscow and other cities.

2. Mark Saul, *Kerosinka: An episode in the history of Soviet mathematics*, Notices of the American Mathematical Society, vol. 46, November 1999, pp. 1217–1220. Available online at http://www.ams.org/notices/199910/fea-saul.pdf

3. George G. Szpiro, *Bella Abramovna Subbotovskaya and the "Jewish People's University,"* Notices of the American Mathematical Society, vol. 54, November 2007, pp. 1326–1330. Available online at http://www.ams.org/notices/200710/tx071001326p.pdf

4. Alexander Shen gives a list of some of the problems that were given to Jewish students at the MGU entrance exams in his article *Entrance examinations to the Mekh-Mat*, Mathematical Intelligencer, vol. 16, No. 4, 1994, pp. 6–10. This article is reprinted in the book M. Shifman (ed.), *You Failed Your Math Test, Comrade Einstein*, World Scientific, 2005 (available online at http://www.ftpi.umn.edu/shifman/ComradeEinstein.pdf). See also other articles about MGU admissions in this book, especially those by I. Vardi and A. Vershik.

Another list of problems is compiled in T. Khovanova and A. Radul, *Jewish Problems*, available at http://arxiv.org/abs/1110.1556

5. George G. Szpiro, ibid.

Chapter 4. Kerosinka

1. Mark Saul, ibid.

Chapter 5. Threads of the Solution

1. The story of the Jewish People's University and the circumstances of Bella Muchnik Subbotovskaya's death are recounted in the articles by D.B. Fuchs and others in M. Shifman (ed.), *You Failed Your Math Test, Comrade Einstein*, World Scientific, 2005.

See also George G. Szpiro, ibid.

2. If we put the identity braid on top of another braid and remove the middle plates, we will get back the original braid after shortening the threads. This means that the result of the addition of a braid b and the identity braid is the same braid b.

3. Here is what the addition of a braid and its mirror image looks like:

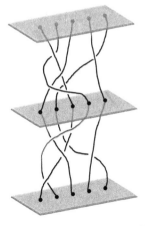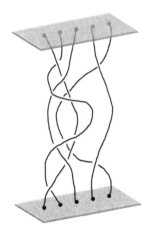

Now, in the braid shown on the right-hand side of the above picture, we pull to the right the thread starting and ending at the right-most "nail." This gives us the left braid below. Then we

do the same with the thread starting and ending at the third nail on this braid. This gives us the right braid below.

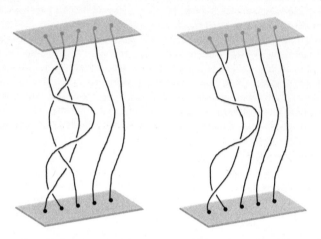

Next, we pull to the left the thread starting and ending on the second nail. In the resulting braid, there is a seeming overlap between the first and the second threads. But this is an illusion: by pulling the second thread to the right, we eliminate this overlap. These moves are shown on the next picture. The resulting braid, on the right-hand side of the picture below, is nothing but the identity braid that we saw above. More precisely, to get the identity braid, we need to straighten the threads, but this is allowed by our rules (we should also shorten the threads, so that our braid has the same height as the original braids). Note that at no step did we cut or sew the threads or allow one to go through the other.

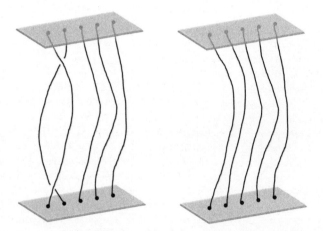

4. This is a good opportunity to discuss the difference between "definition" and "theorem." In Chapter 2, we gave the definition of a group. Namely, a group is a set endowed with an operation (variably called composition, addition, or multiplication, depending on the circumstances) that satisfies the following properties (or axioms): there is an identity element in the set (in the sense explained in Chapter 2); every element of the set has an inverse; and the operation satisfies the associativity property described in endnote 4 to Chapter 2. Once we have given this *definition*, the notion of a group is fixed once and for all. We are not allowed to make any changes to it.

Now, given a set, we can try to endow it with the structure of a group. This means constructing an operation on this set and proving that this operation satisfies all properties listed above. In this chapter, we take the set of all braids with n threads (we identify the braids that are obtained by tweaking the threads, as explained in the main text), and we construct the operation of addition of any two such braids by the rule described in the main text. Our *theorem* is then the statement that this operation satisfies all of the above properties. The proof of this theorem consists of a direct verification of these properties. We have checked the first two properties (see endnotes 2 and 3 above, respectively), and the last property (associativity) follows automatically from the construction of addition of two braids.

5. Because one of our rules is that a thread is not allowed to get entangled with itself, the sole thread we have has no place to go but straight down from the only nail on the top plate to the one on the bottom plate. Of course, it could go along a complicated path, like a winding mountain road or a meandering street, but by shortening it, if needed, we can make the thread go down vertically. In other words, the group B_1 consists of only one element, which is the identity (it is also its own inverse and the result of addition with itself).

6. In mathematical jargon, we say "the braid group B_2 is isomorphic to the group of integers." This means that there is a one-to-one correspondence between the two groups – namely, we assign to each braid the number of overlaps – so that the addition of braids (in the sense described above) corresponds to the usual addition of integers. Indeed, putting two braids on top of each other, we are getting a new braid in which the number of overlaps is equal to the sum of these numbers assigned to the two original braids. Furthermore, the identity braid, in which no overlapping of threads occurs, corresponds to the integer 0, and taking the inverse braid corresponds to taking the negative of an integer.

7. See David Garber, *Braid group cryptography*, in *Braids: Introductory Lectures on Braids, Configurations and Their Applications*, eds. A. Jon Berrick, e.a., pp. 329–403, World Scientific 2010; available at http://arxiv.org/pdf/0711.3941v2.pdf

8. See, for example, Graham P. Collins, *Computing with Quantum Knots*, Scientific American, April 2006, pp. 57–63.

9. De Witt Sumners, Claus Ernst, Sylvia J. Spengler, and Nicholas R. Cozzarelli, *Analysis of the mechanism of DNA recombination using tangles*, Quarterly Reviews of Biophysics, vol. 28, August 1995, pp. 253–313.

Mariel Vazquez and De Witt Sumners, *Tangle analysis of Gin recombination*, Mathematical Proceedings of the Cambridge Philosophical Society, vol. 136, 2004, pp. 565–582.

10. A more precise statement, which we will discuss in Chapter 9, is that the braid group B_n is the *fundamental group* of the space of n distinct unordered points on the plane. Here is a useful interpretation of the collections of n distinct unordered points on the plane in terms of polynomials of degree n. Consider a monic quadratic polynomial $x^2 + a_1 x + a_0$, where a_0 and a_1 are complex numbers ("monic" means here that the coefficient in front of the term with the highest power of x, that is, x^2, is equal to 1). It has two roots, which are complex numbers, and conversely, these roots uniquely determine a monic quadratic polynomial. Complex numbers may be represented as points on the plane (see Chapter 9), so a monic quadratic polynomial with two distinct roots is the same as a pair of distinct points on the plane.

Likewise, a monic polynomial of degree n, $x^n + a_{n-1}x^{n-1} + \ldots + a_1 x + a_0$, with n distinct complex roots is the same as a collection of n distinct points on the plane – its roots. Let's fix one such polynomial: $(x-1)(x-2)\ldots(x-n)$, with the roots $1, 2, 3, \ldots, n$. A path in the space of all such polynomials, starting and ending at the polynomial $(x-1)(x-2)\ldots(x-n)$, may be visualized as a braid with n threads, each thread being the trajectory of a particular root. Hence we find that the braid group B_n is the fundamental group of the space of polynomials of degree n with distinct roots (see Chapter 14).

11. To each overlap between two threads, we assign +1 if the thread coming down from the left goes under the thread coming down from the right; we assign -1 if the opposite is true. Consider, for example, this braid:

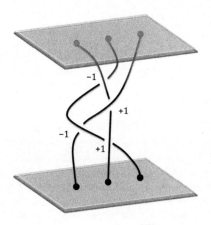

When we sum up these numbers (+1 and -1) over all pairwise overlaps, we obtain the total overlap number of a given braid. If we tweak the threads, we will always add or eliminate the same number of +1 overlaps as the number of -1 overlaps, so the total overlap number will stay the same. This means that the total overlap number is *well-defined*: it does not change when we tweak the braid.

12. Note that the total overlap number of the braid obtained by the addition of two braids will be equal to the sum of the total overlap numbers of those two braids. Therefore, the addition of two braids having total overlap numbers 0 will again be a braid with the total overlap number 0. The commutator subgroup B'_n consists of all such braids. In a certain precise sense, it is the maximal non-abelian part of the braid group B_n.

13. The concept of Betti numbers originated in topology, the mathematical study of the salient properties of geometric shapes. The Betti numbers of a given geometric shape, such as a circle or a sphere, form a sequence of numbers, $b_0, b_1, b_2, \ldots$, each of which could be either 0 or a natural number. For example, for a flat space, such as a line, a plane, etc., $b_0 = 1$, and all other Betti numbers are equal to 0. In general, b_0 is the number of connected components of the geometric shape. For the circle, $b_0 = 1, b_1 = 1$, and the rest of the Betti numbers are 0. The fact that b_1, the first Betti number, is equal to 1 reflects the presence of a non-trivial one-dimensional piece. For the sphere, $b_0 = 1, b_1 = 0, b_2 = 1$, and the other Betti numbers are all equal to 0. Here b_2 reflects the presence of a non-trivial two-dimensional piece.

The Betti numbers of the braid group B_n are defined as the Betti numbers of the space of monic polynomials of degree n with n distinct roots. The Betti numbers of the commutator subgroup B_n' are the Betti numbers of a closely related space. It consists of all monic polynomials of degree n with n distinct roots and with the additional property that their discriminant (the square of the product of the differences between all pairs of roots) takes a fixed non-zero value (for instance, we can say that this value is 1). For example, the discriminant of the polynomial $x^2 + a_1 x + a_0$ is equal to $a_1^2 - 4a_0$, and there is a similar formula for all n.

It follows from the definition that the discriminant of a polynomial is equal to zero if and only if it has multiple roots. Therefore, the discriminant gives us a map from the space of all monic polynomials of degree n with n distinct roots to the complex plane without the point 0. Thus, we obtain a "fibration" of this space over the complex plane without the origin. The Betti numbers of B_n' reflect the topology of any of these fibers (topologically, they are the same), while the Betti numbers of B_n reflect the topology of the entire space. The desire to understand the topology of the fibers was what motivated Varchenko to suggest this problem to me in the first place. For more on the Betti numbers and the related concepts of homology and cohomology, you may consult the following introductory textbooks:

William Fulton, *Algebraic Topology: A First Course*, Springer, 1995;

Allen Hatcher, *Algebraic Topology*, Cambridge University Press, 2001.

Chapter 6. Apprentice Mathematician

1. Some have speculated that Fermat may have bluffed when he left that note on the margin. I don't think so; I think he made an honest mistake. Anyway, we have to be grateful to him – his little note on the margin has definitely had a positive effect on the development of mathematics.

2. More precisely, I proved that for each divisor d of n, the qth Betti number, where $q = n(d - 2)/d$, is equal to $\phi(d)$, and for each divisor d of $n - 1$, the qth Betti number, where $q = (n - 1)(d - 2)/d$, is equal to $\phi(d)$. All other Betti numbers of B_n' are equal to 0.

3. In 1985, Mikhail Gorbachev came to power, and soon afterward he launched his policy of *perestroika*. As far as I know, systematic discrimination of Jewish applicants at the entrance exams to *Mekh-Mat* of the kind that I had experienced ended around 1990.

4. S. Zdravkovska and P. Duren, *Golden Years of Moscow Mathematics*, American Mathematical Society, 1993, p. 221.

5. Mathematician Yuly Ilyashenko argued that this event was a catalyst for the establishment of anti-Semitic policies at *Mekh-Mat* in the interview entitled *The black 20 years at Mekh-Mat*, published on the website Polit.ru on July 28, 2009:

http://www.polit.ru/article/2009/07/28/ilyashenko2

6. The question was to find in how many ways can one glue pairwise the sides of a regular polygon with $4n$ sides to obtain a Riemann surface of genus n. In Chapter 9, we will discuss a particular way of doing so when we identify the opposite sides of the polygon.

7. Edward Frenkel, *Cohomology of the commutator subgroup of the braid group*, Functional Analysis and Applications, vol. 22, 1988, pp. 248–250.

Chapter 7. The Grand Unified Theory

1. Interview with Robert Langlands for the Mathematics Newsletter, University of British Columbia (2010), full version available at

http://www.math.ubc.ca/Dept/Newsletters/Robert_Langlands_interview_2010.pdf

2. Suppose that there exist natural numbers m and n such that $\sqrt{2} = m/n$. We can assume without loss of generality that the numbers m and n are relatively prime; that is, they are not simultaneously divisible by any natural number other than 1. Otherwise, we would have $m = dm'$ and $n = dn'$ and then $\sqrt{2} = m'/n'$. This process can be repeated, if needed, until we reach two numbers that are relatively prime.

So let us assume that $\sqrt{2} = m/n$, where m and n are relatively prime. Squaring both sides of the formula $\sqrt{2} = m/n$, we obtain $2 = m^2/n^2$. Multiplying both sides by n^2, we have $m^2 = 2n^2$. This implies that m is even, for if it were odd, then m^2 would also be odd, which would contradict this formula.

If m is even, then $m = 2p$ for some natural number p. Substituting this into the previous formula, we obtain $4p^2 = 2n^2$, hence $n^2 = 2p^2$. But then n must also be even, by the same argument as the one we used to show that m is even. Thus, both m and n must be even, which contradicts our assumption that m and n are relatively prime. Hence, such m and n do not exist.

This is a good example of a "proof by contradiction." We start with the statement that is opposite to what we are trying to prove (in our case, we start with the statement that $\sqrt{2}$ is a rational number, which is opposite to what we are trying to prove). If this implies a false statement (in our case, this implies that both m and n are even, even though we had assumed that they were relatively prime), then we conclude that the statement we started with is also false. Hence the statement we wanted to prove (that $\sqrt{2}$ is not a rational number) is true. We will use this method again in Chapter 8: first, when we discuss the proof of Fermat's Last Theorem, and then again, in endnote 6, when we give Euclid's proof that there are infinitely many prime numbers.

3. For example, let us multiply these two numbers: $\frac{1}{2} + \sqrt{2}$ and $3 - \sqrt{2}$. We simply open the brackets:

$$\left(\frac{1}{2} + \sqrt{2}\right)(3 - \sqrt{2}) = \frac{1}{2} \cdot 3 - \frac{1}{2} \cdot \sqrt{2} + \sqrt{2} \cdot 3 - \sqrt{2} \cdot \sqrt{2}.$$

But $\sqrt{2} \cdot \sqrt{2} = 2$, so by collecting the terms, we obtain the following answer:

$$\frac{3}{2} - \frac{1}{2}\sqrt{2} + 3\sqrt{2} - 2 = -\frac{1}{2} + \frac{5}{2}\sqrt{2}.$$

It is a number of the same form, so it does belong to our new numerical system.

4. We only consider the symmetries of our numerical system that are compatible with the operations of addition and multiplication, and such that 0 goes to 0, 1 goes to 1, additive inverse goes to additive inverse, and multiplicative inverse goes to multiplicative inverse. But if 1 goes to 1, then $2 = 1 + 1$ must go to $1 + 1 = 2$. Likewise, all natural numbers must be preserved, and then so are their negatives and multiplicative inverses. Hence, all rational numbers are preserved by such symmetries.

5. It is easy to check that this symmetry is indeed compatible with addition, subtraction, multiplication, and division. Let's do this for the operation of addition. Consider two different numbers in our new numerical system:

$$x + y\sqrt{2} \qquad \text{and} \qquad x' + y'\sqrt{2},$$

where x, y, x', y, are rational numbers. Let's add them:

$$\left(x + y\sqrt{2}\right) + \left(x' + y'\sqrt{2}\right) = (x + x') + (y + y')\sqrt{2}.$$

We can apply our symmetry to each of them. We then get:

$$x - y\sqrt{2} \quad \text{and} \quad x' - y'\sqrt{2},$$

Now let's add these:

$$\left(x - y\sqrt{2}\right) + \left(x' - y'\sqrt{2}\right) = \left(x + x'\right) - \left(y + y'\right)\sqrt{2}.$$

We see that the number we get is equal to the number obtained by applying our symmetry to the original sum

$$\left(x + x'\right) + \left(y + y'\right)\sqrt{2} \quad \mapsto \quad \left(x + x'\right) - \left(y + y'\right)\sqrt{2}.$$

In other words, we can apply the symmetry to each of the two numbers individually and then add them up. Or we can first add them up, and then apply the symmetry. The result will be the same. This is what we mean by saying that our symmetry is *compatible* with the operation of addition. Likewise, we can check that our symmetry is compatible with the operations of subtraction, multiplication, and division.

6. For example, in the case of the number field obtained by adjoining $\sqrt{2}$ to the rational numbers, the Galois group consists of two symmetries: the identity and the symmetry exchanging $\sqrt{2}$ and $-\sqrt{2}$. Denote the identity by I and the symmetry exchanging $\sqrt{2}$ and $-\sqrt{2}$ by S. Let's write down explicitly what the compositions of these symmetries are:

$$I \circ I = I, \quad I \circ S = S, \quad S \circ I = S,$$

and the most interesting one:

$$S \circ S = I.$$

Indeed, if we exchange $\sqrt{2}$ and $-\sqrt{2}$ and then do this again, the net result will be the identity:

$$x + y\sqrt{2} \quad \mapsto \quad x - y\sqrt{2} \quad \mapsto \quad x - (-y\sqrt{2}) = x + y\sqrt{2}.$$

We have now completely described the Galois group of this number field: it consists of two elements, I and S, and their compositions are given by the above formulas.

7. A few years earlier, Niels Henrik Abel showed that there was a quintic equation that could not be solved in radicals (Joseph-Louis Lagrange and Paolo Ruffini also made important contributions). However, Galois' proof was more general and more conceptual. For more on the Galois groups and the rich history of solving polynomial equations, see Mario Livio, *The Equation That Couldn't Be Solved*, Simon & Schuster, 2005.

8. More generally, consider the quadratic equation $ax^2 + bx + c = 0$ with rational coefficients a, b, c. Its solutions x_1 and x_2 are given by the formulas

$$x_1 = \frac{-b + \sqrt{b^2 - 4ac}}{2a} \quad \text{and} \quad x_2 = \frac{-b - \sqrt{b^2 - 4ac}}{2a}.$$

If the discriminant $b^2 - 4ac$ is not the square of a rational number, then these solutions are not rational numbers. Hence, if we adjoin x_1 and x_2 to the rational numbers, we obtain a new number field. The group of symmetries of this number field also consists of two elements: the identity and the symmetry exchanging the two solutions, x_1 and x_2. In other words, this symmetry exchanges $\sqrt{b^2 - 4ac}$ and $-\sqrt{b^2 - 4ac}$.

But we don't need to write down explicit formulas for the solutions to describe this Galois group. Indeed, because the degree of the polynomial is two, we know that there are two solutions, so let's just denote them by x_1 and x_2. Then we have

$$ax^2 + bx + c = a(x - x_1)(x - x_2).$$

Opening the brackets, we find that $x_1 + x_2 = -\dfrac{b}{a}$, so that $x_2 = -\dfrac{b}{a} - x_1$. We also have $(x_1)^2 = -\dfrac{c + bx_1}{a}$ because x_1 is a solution of the above equation. Therefore, if the discriminant is not the square of a rational number, then the number field obtained by adjoining x_1 and x_2 to the rational numbers consists of all numbers of the form $\alpha + \beta x_1$, where α and β are two rational numbers. Under the symmetry exchanging x_1 and x_2, the number $\alpha + \beta x_1$ goes to

$$\alpha + \beta x_2 = \left(\alpha - \beta \frac{b}{a}\right) - \beta x_1.$$

This symmetry is compatible with the operations of addition, etc., because both x_1 and x_2 solve the same equation with rational coefficients. We obtain that the Galois group of this number field consists of the identity and the symmetry exchanging x_1 and x_2. I stress again that we did not use any knowledge whatsoever about how to express x_1 and x_2 in terms of a, b, c.

9. To illustrate this point, consider, for example, the equation $x^3 = 2$. One of its solutions is the cubic root of 2, $\sqrt[3]{2}$. There are two more solutions, which are complex numbers: $\sqrt[3]{2}\,\omega$ and $\sqrt[3]{2}\,\omega^2$, where

$$\omega = -\frac{1}{2} + \frac{\sqrt{3}}{2}\sqrt{-1}$$

(see the discussion of complex numbers in Chapter 9). The smallest number field containing these three solutions should also contain their squares: $\sqrt[3]{4} = (\sqrt[3]{2})^2$, $\sqrt[3]{4}\,\omega$, and $\sqrt[3]{4}\,\omega^2$ (their cubes are equal to 2), as well as their ratios: ω and ω^2. So it looks like to construct this number field, we have to adjoin eight numbers to the rationals. However, we have a relation:

$$1 + \omega + \omega^2 = 0,$$

which allows us to express ω^2 in terms of 1 and ω:

$$\omega^2 = -1 - \omega.$$

Hence we also have

$$\sqrt[3]{2}\,\omega^2 = -\sqrt[3]{2} - \sqrt[3]{2}\,\omega, \qquad \sqrt[3]{4}\,\omega^2 = -\sqrt[3]{4} - \sqrt[3]{4}\,\omega.$$

Therefore to obtain our number field, we only need to adjoin five numbers to the rationals: $\omega, \sqrt[3]{2}, \sqrt[3]{2}\,\omega, \sqrt[3]{4}$, and $\sqrt[3]{4}\,\omega$. Hence a general element of this number field, called the splitting field of the equation $x^3 = 2$, will be a combination of six terms: a rational number plus a rational number times ω plus a rational number times $\sqrt[3]{2}$, and so on. Compare this with the splitting field of the equation $x^2 = 2$, whose elements have two terms: a rational number plus a rational number times $\sqrt{2}$.

We have seen above that the elements of the Galois group of the splitting field of the equation $x^2 = 2$ permute the two solutions of this equation, $\sqrt{2}$ and $-\sqrt{2}$. There are two such permutations: the one switching these two solutions and the identity.

Likewise, for any other equation with rational coefficients, we define its splitting field as the field obtained by adjoining all of its solutions to the rational numbers. By the same argument as in endnote 4 above, any symmetry of this number field compatible with the operations of addition and multiplication preserves the rational numbers. Therefore, under such a symmetry, any solution of this equation must go to another solution. Hence, we obtain a permutation of these solutions. In the case of the equation $x^3 = 2$, there are three solutions listed above. Under each permutation, the first, $\sqrt[3]{2}$, goes to any one of the three solutions; the second, $\sqrt[3]{2}\,\omega$, goes to one of the remaining two solutions; and then the third, $\sqrt[3]{2}\,\omega^2$, must go to one remaining solution (a permutation must be one-to-one in order to have an inverse). Therefore, there are $3 \cdot 2 = 6$ possible permutations of these three solutions. These permutations form a group, and it turns out that this group is in a one-to-one correspondence with the Galois group of the splitting field of the equation $x^3 = 2$. Thus, we obtain an explicit description of the Galois group in terms of permutation of the solutions.

In the above calculation, we used explicit formulas for the solutions of the equation. But a similar argument can be made for an arbitrary cubic equation with rational coefficients, and we don't need a formula for its solutions in terms of the coefficients. The result is the following: let us denote the solutions of the equation by x_1, x_2, and x_3. Assume that all of them are irrational. However, it is easy to see that the discriminant of the equation, defined as

$$(x_1 - x_2)^2 (x_1 - x_3)^2 (x_2 - x_3)^2,$$

is always a rational number. It turns out that if its square root is not a rational number, then the Galois group of the splitting field of this equation is the group of all permutations of these solutions (it then consists of six elements). If the the square root of the discriminant is a rational number, then the Galois group consists of three permutations: the identity, the cyclic permutation $x_1 \mapsto x_2 \mapsto x_3 \mapsto x_1$, and its inverse.

10. For example, it is not difficult to show that for a typical quintic equation (that is, one with $n = 5$), for which we have five solutions, the Galois group is the group of all permutations of these five numbers. A permutation is a one-to-one reshuffling of these numbers, such as the one shown on the picture below.

Under such a permutation, the solution x_1 goes to any of the five (possibly, to itself), so we have five choices, then x_2 has to go to one of the remaining four solutions, x_3 to one of the remaining three, and so on. Hence, altogether, there are $5 \cdot 4 \cdot 3 \cdot 2 \cdot 1 = 120$ permutations, and so the Galois group consists of 120 elements.

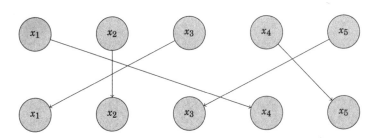

(The group of permutations of a set of n elements, also known as the symmetric group on n letters, consists of $n! = n \cdot (n-1) \cdot \ldots 2 \cdot 1$ elements.) Unlike the Galois groups of the quadratic, cubic, and quartic equations, it is not a solvable group. Therefore, according to Galois' argument, we cannot express solutions of the general quintic equation in terms of radicals.

11. It is now available on the website of the Institute for Advanced Study in Princeton, http://publications.ias.edu/sites/default/files/weil1.pdf

12. Quoted from the image available at the Institute for Advanced Study Digital Collections, http://cdm.itg.ias.edu/cdm/compoundobject/collection/coll12/id/1682/rec/1

Chapter 8. Magic Numbers

1. Robert Langlands, *Is there beauty in mathematical theories?*, in *The Many Faces of Beauty*, ed. Vittorio Hösle, University of Notre Dame Press, 2013, available online at

http://publications.ias.edu/sites/default/files/ND.pdf

2. For more on conjectures, see this insightful article: Barry Mazur, *Conjecture*, Synthèse, vol. 111, 1997, pp. 197–210.

3. For more on the history of Fermat's Last Theorem, see Simon Singh, *Fermat's Enigma: The Epic Quest to Solve the World's Greatest Mathematical Problem*, Anchor, 1998.

4. See Andrew Wiles, *Modular elliptic curves and Fermat's last theorem*, Annals of Mathematics, vol. 141, 1995, pp. 443–551;

Richard Taylor and Andrew Wiles, *Ring-theoretic properties of certain Hecke algebras*, Annals of Mathematics, vol. 141, 1995, 553–572.

They proved the Shimura–Taniyama–Weil conjecture in the most typical (the so-called semistable) case, which turned out to be sufficient to settle Fermat's Last Theorem. A few years later, the remaining cases of the conjecture were proved by C. Breuil, B. Conrad, F. Diamond, and R. Taylor.

Because it is now proved, it would be more proper to refer to the Shimura–Taniyama–Weil conjecture as a theorem. And in fact, many mathematicians now refer to it as the "modularity theorem." But old habits die hard, and some, like me, still use its old name. Ironically, Fermat's Last Theorem has always been referred to as a theorem, even though it was in fact a conjecture. No doubt, initially this was done out of respect for Fermat's claim that he had found a proof.

5. If N is not a prime, then we can write $N = xy$ for two natural numbers x and y between 1 and $N-1$. Then x does not have a multiplicative inverse modulo N. In other words, there does not exist a natural number z between 1 and $N-1$ such that

$$xz = 1 \qquad \text{modulo} \quad N.$$

Indeed, if this equality were satisfied, we would multiply both sides by y, and we would obtain

$$xyz = y \qquad \text{modulo} \quad N.$$

But $xy = N$, so the left-hand side is Nz, which means that y is divisible by N. But then y cannot be between 1 and $N-1$.

6. The proof attributed to Euclid goes as follows. We apply the method of "proof by contradiction," which we have already used in this chapter when we discussed the proof of Fermat's Last Theorem.

Suppose that there are only finitely many primes: $p_1, p_2, \ldots, p_N$. Consider the number A obtained by taking their product and adding 1; that is, set $A = p_1 p_2 \ldots p_N + 1$. I claim that it is a prime number. We prove this by contradiction: if it is not a prime number, then it is divisible by a natural number other than 1 and itself. Hence A has to be divisible by one of the prime numbers – let's say, by p_i. But if A is divisible by p_i, then $A = 0$ modulo p_i, whereas it follows from the definition of A that $A = 1$ modulo p_i.

We have arrived at a contradiction. This means that A is not divisible by any natural number other than 1 and itself. Hence A is itself a prime number.

But since A is clearly larger than any of the numbers $p_1, p_2, \ldots, p_N$, this contradicts our assumption that $p_1, p_2, \ldots, p_N$ were the only prime numbers. Therefore our initial statement that there are only finitely many primes is false. Hence, there are infinitely many prime numbers.

7. Let's spell this out: within a particular numerical system, a multiplicative inverse of a number a is a number b such that $a \cdot b = 1$. So, for example, within the numerical system of rational numbers, the multiplicative inverse of the rational number $\frac{3}{4}$ is $\frac{4}{3}$. Within the numerical system we are considering now, the inverse of a natural number a between 1 and $p-1$ is another natural number b in the same range such that

$$a \cdot b = 1 \qquad \text{modulo} \quad p.$$

No matter which numerical system we consider, the number 0, the additive identity, never has a multiplicative inverse. That's why we exclude it.

8. Here is the proof. Let us pick a natural number a between 1 and $p-1$, where p is a prime number. Let's multiply a by all other numbers b in this range and take the result modulo p. We will compile a table consisting of two columns: in the first column will be the number b, and in the second column will be the number $a \cdot b$ modulo p.

For instance, if $p = 5$ and $a = 2$, this table looks as follows:

1	2
2	4
3	1
4	3

We see right away that each of the numbers $1, 2, 3, 4$ appears in the right column exactly once. What happens when we multiply by 2 is that we obtain the same set of numbers, but they are permuted in a certain way. In particular, number 1 appears in the third line. This means that when we multiply 3 by 2, we get 1 modulo 5. In other words, 3 is the inverse of 2 if we do arithmetic modulo 5.

The same phenomenon holds in general: if we compile a table like the one above for any prime p and any number a from the list $1, 2, \ldots, p-1$, then each of the numbers $1, 2, \ldots, p-1$ will appear in the right column exactly once.

Let us prove this, again employing the trick of proof by contradiction: suppose that this is not the case. Then one of the numbers from the set $1, 2, \ldots, p-1$, call it n, has to appear in the right column at least two times. This means that there are two numbers from the set $1, 2, \ldots, p-1$, call them c_1 and c_2 (suppose that $c_1 > c_2$), such that

$$a \cdot c_1 = a \cdot c_2 = n \quad \text{modulo} \quad p.$$

But then we have

$$a \cdot c_1 - a \cdot c_2 = a \cdot (c_1 - c_2) = 0 \quad \text{modulo} \quad p.$$

The last formula means that $a \cdot (c_1 - c_2)$ is divisible by p. But this is impossible because p is a prime and both a and $c_1 - c_2$ are from the set $\{1, 2, ..., p-1\}$.

We conclude that in the right column of our table each of the numbers $1, 2, ..., p-1$ appears no more than once. But because there are exactly $p-1$ of those numbers and we have the same number of lines in our table, $p-1$, the only possibility for this to happen is for each number to appear exactly once. But then number 1 has to appear somewhere in the right column and exactly once. Let b be the corresponding number in the left column. But then we have

$$a \cdot b = 1 \quad \text{modulo} \quad p.$$

This completes the proof.

9. For example, we can divide 4 by 3 in the finite field of 5 elements:

$$4/3 = 4 \cdot 3^{-1} = 4 \cdot 2 = 8 \quad \text{modulo} \quad 5$$
$$= 3 \quad \text{modulo} \quad 5$$

(here we use the fact that 2 is the multiplicative inverse of 3 modulo 5).

10. Let us observe that for any number a whose absolute value is less than 1 we have

$$1 + a + a^2 + a^3 + a^4 + \ldots = \frac{1}{1-a},$$

which is easy to prove by multiplying both sides by $1 - a$. Using this identity, and denoting $(q + q^2)$ by a, we can rewrite the generating function for the Fibonacci numbers

$$q\left(1 + (q + q^2) + (q + q^2)^2 + (q + q^2)^3 + \ldots\right)$$

as

$$\frac{q}{1 - q - q^2}.$$

Next, writing $1 - q - q^2$ as a product of linear factors, we find that

$$\frac{q}{1 - q - q^2} = \frac{1}{\sqrt{5}}\left(\left(1 - \frac{1 + \sqrt{5}}{2}q\right)^{-1} - \left(1 - \frac{1 - \sqrt{5}}{2}q\right)^{-1}\right).$$

Using the above identity again, with $a = \dfrac{1 \pm \sqrt{5}}{2}q$, we find that the coefficient in front of q^n in our generating function (which is F_n) is equal to

$$F_n = \frac{1}{\sqrt{5}}\left(\left(\frac{1 + \sqrt{5}}{2}\right)^n - \left(\frac{1 - \sqrt{5}}{2}\right)^n\right).$$

We therefore obtain a closed formula for the nth Fibonacci number, which is independent of the preceding ones.

Note that the number $\dfrac{1 + \sqrt{5}}{2}$, which appears in this formula, is the so-called *golden ratio*. It follows from the above formula that the ratio F_n/F_{n-1} tends to the golden ratio as n becomes larger. For more on the golden ratio and the Fibonacci numbers, see Mario Livio, *The Golden Ratio*, Broadway, 2003.

11. I follow the presentation of this result given in Richard Taylor, *Modular arithmetic: driven by inherent beauty and human curiosity*, The Letter of the Institute for Advanced Study,

Summer 2012, pp. 6–8. I thank Ken Ribet for useful comments. According to André Weil's book *Dirichlet Series and Automorphic Forms*, Springer-Verlag, 1971, the cubic equation we are discussing in this chapter was introduced by John Tate, following Robert Fricke.

12. This group is one of the so-called "congruence subgroups" of the group called $SL_2(\mathbb{Z})$, which consists of 2×2 matrices with integer coefficients and with the determinant 1, that is, the arrays of integers

$$\begin{pmatrix} a & b \\ c & d \end{pmatrix}$$

such that $ad - bc = 1$. The multiplication of matrices is given by the standard formula

$$\begin{pmatrix} a & b \\ c & d \end{pmatrix} \cdot \begin{pmatrix} a' & b' \\ c' & d' \end{pmatrix} = \begin{pmatrix} aa' + bc' & ab' + bd' \\ ca' + dc' & cb' + dd' \end{pmatrix}$$

Now, any complex number q inside the unit disc may be written as $e^{2\pi\tau\sqrt{-1}}$ for some complex number τ whose imaginary part is positive: $\tau = x + y\sqrt{-1}$, where $y > 0$ (see endnote 12 to Chapter 15). The number q is uniquely determined by τ, and vice versa. Hence, we can describe the action of the group $SL_2(\mathbb{Z})$ on q by describing the corresponding action on τ. The latter is given by the following formula:

$$\begin{pmatrix} a & b \\ c & d \end{pmatrix} \cdot \tau = \frac{a\tau + b}{c\tau + d} .$$

The group $SL_2(\mathbb{Z})$ (more precisely, its quotient by the two-element subgroup that consists of the identity matrix I and the matrix $-I$) is the group of symmetries of the disc endowed with a particular non-Euclidean metric, which is called the Poincaré disc model. Our function is a modular form of "weight 2," which means that it is invariant under the above action of a congruence subgroup of $SL_2(\mathbb{Z})$ on the disc, if we correct this action by multiplying the function by the factor $(c\tau + d)^2$.

See, for example, Henri Darmon, *A proof of the full Shimura–Taniyama–Weil conjecture is announced*, Notices of the American Mathematical Society, vol. 46, December 1999, pp. 1397–1401. Available online at http://www.ams.org/notices/199911/comm-darmon.pdf

13. This picture was created by Lars Madsen and is published with his permission. I thank Ian Agol for pointing it out to me and a useful discussion.

14. See, for example, Neal Koblitz, *Elliptic curve cryptosystems*, Mathematics of Computation, vol. 49, 1987, pp. 203–209;

I. Blake, G. Seroussi, and N. Smart, *Elliptic Curves in Cryptography*, Cambridge University Press, 1999.

15. In general, this is true for all but finitely many primes p. There is also an additional pair of invariants attached to the cubic equation (the so-called conductor) and to the modular form (the so-called level), and these invariants are also preserved under this correspondence. For example, in the case of the cubic equation we have considered, they are both equal to 11. I also note that every modular forms that appears here has zero constant term, the coefficient b_1 in front of q is equal to 1, and all other coefficients b_n with $n > 1$ are determined by the b_p corresponding to the primes p.

16. Namely, if a, b, c solve the Fermat equation $a^n + b^n = c^n$, where n is an odd prime number, then consider, following Yves Hellegouarch and Gerhard Frey, the cubic equation

$$y^2 = x(x - a^n)(x + b^n).$$

Ken Ribet proved (following a suggestion of Frey and some partial results obtained by Jean-Pierre Serre) that this equation cannot satisfy the Shimura–Taniyama–Weil conjecture. Together with the case $n = 4$ (which was in fact proved by Fermat himself), this implies Fermat's Last Theorem. Indeed, any integer $n > 2$ may be written as a product $n = mk$, where m is either 4 or an odd prime. Therefore the absence of solutions to the Fermat equation for such m implies their absence for all $n > 2$.

17. Goro Shimura, *Yutaka Taniyama and his time. Very personal recollections*, Bulletin of London Mathematical Society, vol. 21, 1989, p. 193.

18. Ibid., p. 190.

19. See footnote 1 on pp. 1302–1303 in the following article on the rich history of the conjecture: Serge Lang, *Some history of the Shimura–Taniyama conjecture*, Notices of the American Mathematical Society, vol. 42, 1995, pp. 1301–1307. Available online at

http://www.ams.org/notices/199511/forum.pdf

Chapter 9. Rosetta Stone

1. *The Economist*, August 20, 1998, p. 70.

2. The pictures of Riemann surfaces in this book were created with *Mathematica®* software, using the code kindly provided by Stan Wagon. For more detail, see his book: Stan Wagon, *Mathematica® in Action: Problem Solving Through Visualization and Computation*, Springer-Verlag, 2010.

3. This is not a precise definition, but it gives the right intuition for real numbers. To get a precise definition, we should think of every real number as the limit of a converging sequence of rational numbers (also known as a Cauchy sequence); for example, the truncations of the infinite decimal expansion of $\sqrt{2}$ yield such a sequence.

4. In order to do this, mark one point on the circle and put the circle on the line, so that this marked point touches point 0 on the line. Then roll the circle to the right until the marked point on the circle again touches the line (this will happen after the circle makes one full turn). This point of contact between the circle and the line will be the point corresponding to π.

5. The geometry of complex numbers (and other numerical systems) is beautifully explained in Barry Mazur, *Imagining Numbers*, Picador, 2004.

6. More precisely, we get the surface of the donut without one point. This extra point corresponds to the "infinite solution," when both x and y tend to infinity.

7. To get a Riemann surface of genus g, we should put a polynomial in x of degree $2g + 1$ on the right-hand side of the equation.

8. This link between algebra and geometry was a profound insight of René Descartes, first described in *La Géométrie*, an appendix to his book *Discours de la Méthode*, published in 1637. Here is what E.T. Bell wrote about Descartes' method: "Now comes the real power of his method. *We start with equations of any desired or suggested degree of complexity and interpret their algebraic and analytic properties geometrically.... Henceforth algebra and analysis are to be our pilots to the uncharted seas of 'space' and its 'geometry.'* " (E.T. Bell, *Men of Mathematics*, Touchstone, 1986, p. 54). Note however that Descartes' method applies to solutions of equations in real numbers, whereas in this chapter we are interested in solutions in finite fields and in complex numbers.

9. For example, we learned in Chapter 8 that the cubic equation $y^2 + y = x^3 - x^2$ has four solutions modulo 5. So, naively, the corresponding curve over the finite field of 5 elements has four points. But in fact there is a lot more structure because we can also consider solutions with values in various extensions of the finite field of 5 elements; for example, the field obtained by adjoining the solutions of the equation $x^2 = 2$, which we discuss in endnote 8 to Chapter 14. These extended fields have 5^n elements for $n = 2, 3, 4, \ldots$, and so we get a hierarchy of solutions with values in these finite fields.

The curves corresponding to the cubic equations are called "elliptic curves."

10. *The Bhagavad-Gita*, Krishna's Counsel in Time of War, translated by Barbara Stoler Miller, Bantam Classic, 1986.

It is interesting to note that Weil spent two years in India in the early 1930s and, by his own admission, was influenced by the Hindu religion.

11. See, for example, Noel Sheth, *Hindu Avatāra and Christian Incarnation: A comparison*, Philosophy East and West, vol. 52, No. 1, pp. 98–125.

12. André Weil, *Collected Papers*, vol. I, Springer-Verlag, 1979, p. 251 (my translation).

13. Ibid., p. 253. The idea is that given a curve over a finite field, we consider the so-called rational functions on it. These functions are ratios of two polynomials. (Note that such a function has a "pole" – that is, its value is undefined – at each point of the curve where the polynomial appearing in the denominator has a zero.) It turns out that the set of all rational functions on a given curve is analogous in its properties to the set of rational numbers, or a more general number field, like the ones we discussed in Chapter 8.

To explain this more precisely, let's consider rational functions on Riemann surfaces; the analogy will still be valid. For example, consider the sphere. Using the stereographic projection, we can view the sphere as a union of a point and the complex plane (we can think of the extra point as one representing infinity). Denote by $t = r + s\sqrt{-1}$ the coordinate on the complex plane. Then each polynomial $P(t)$ with complex coefficients is a function on the plane. These polynomials are the analogues of the integers appearing in number theory. A rational function on the sphere is a ratio of two polynomials $P(t)/Q(t)$ without common factors. These rational functions are the analogues of the rational numbers, which are ratios m/n of integers without common factors. Similarly, rational functions on a more general Riemann surface are analogous to elements of a more general number field.

The power of this analogy lies in the fact that for many results about number fields there will be similar results valid for the rational functions on curves over finite fields, and vice versa. Sometimes, it is easier to spot and/or to prove a particular statement for one of them. Then the analogy would tell us that a similar statement must be true for the other. This has been one of the vehicles used by Weil and other mathematicians to produce new results.

14. Ibid., p. 253. Here I use the translation by Martin H. Krieger in Notices of the American Mathematical Society, vol. 52, 2005, p. 340.

15. There were three Weil conjectures, which were proved by Bernard Dwork, Alexander Grothendieck, and Pierre Deligne.

16. There is a redundancy in this definition. To explain this, consider two paths on the plane shown on the picture below, one solid and one dotted. It is clear that one of them can be continuously deformed into another without breaking. It is reasonable and economical to declare two closed paths that may be deformed into one another in this way as equal. If we do this, we drastically reduce the number of elements in our group.

This rule is in fact similar to the rule we used in the definition of braid groups in Chapter 5. There, too, we declared equal two braids that could be deformed (or "tweaked") into one another without cutting and sewing the threads.

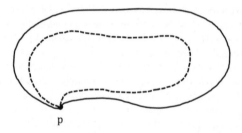

p

So we define the fundamental group of our Riemann surface as the group whose elements are closed paths starting and ending at the point P, with the additional requirement that we identify the paths that can be continuously deformed into one another.

Note that if our Riemann surface is connected, which we tacitly assume to be the case throughout, then the choice of the reference point P is inessential: the fundamental groups assigned to different reference points P will be in one-to-one correspondence with each other (more precisely, they will be "isomorphic" to each other).

17. The identity element will be the "constant path." It never leaves the marked point P. In fact, it is instructive to think of each closed path as a trajectory of a particle, starting and ending at the same point P. The constant path is the trajectory of the particle that just stays at the point P. It is clear that if we add any path to the constant path, in the sense described in the main text, we will get back the original path.

The inverse path to a given path will be the same path, but traversed in the opposite direction. To check that it is indeed the inverse, let us add a path and its inverse. We obtain a new path which traverses the same route twice, but in two opposite directions. We can continuously deform this new "double" path to the constant path. First, we tweak one of the two paths slightly. The resulting path may be contracted to a point, as shown on the pictures below.

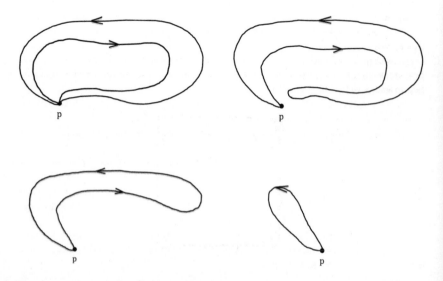

18. Alternatively, as we discussed in endnote 10 to Chapter 5, the braid group B_n may be interpreted as the fundamental group of the space of monic polynomials of degree n with n distinct roots. We choose as the reference point P, the polynomial $(x-1)(x-2)\ldots(x-n)$ with the roots $1, 2, \ldots, n$ (these are the "nails" of the braid).

19. To see that the two paths commute with each other, let's observe that the torus may be obtained by gluing the opposite sides of a square (polygon with 4 vertices). When we glue together two horizontal sides, a_1 and a_1', we obtain a cylinder.

Gluing the circles at the opposite ends of the cylinder (which is what the other two vertical sides of the square, a_2 and a_2', become after the first gluing), we obtain a torus. Now we see that the sides a_1 and a_2 become the two independent closed paths on the torus. Note that on the torus all four corners represent the same point, so these two paths become closed – they start and end at the same point P on the torus. Also, $a_1 = a_1'$ because we have glued them together, and likewise, $a_2 = a_2'$.

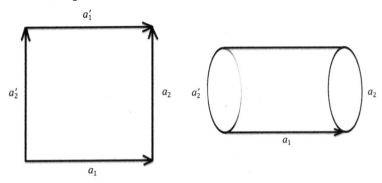

On the square, if we take the path a_1 and then take the path a_2, this will take us from one corner to its opposite. The resulting path is $a_1 + a_2$. But we can also go between these corners along a different path: first take a_2' and then a_1', which is the same as a_1. The resulting path is $a_2' + a_1'$. After gluing the opposite sides of the square, a_1' becomes a_1, and a_2' becomes a_2. So $a_2' + a_1' = a_2 + a_1$.

Now observe that both $a_1 + a_2$ and $a_2 + a_1$ can be deformed to the diagonal path, a straight line connecting the two opposite corners, as shown on the picture below (the dashed arrows show how to deform each of the two paths).

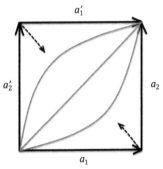

This means that the paths $a_1 + a_2$ and $a_2 + a_1$ give rise to the same element in the fundamental group of the torus. We have proved that

$$a_1 + a_2 = a_2 + a_1.$$

This implies that the fundamental group of the torus has a simple structure: we can express its elements as $M \cdot a_1 + N \cdot a_2$, where a_1 and a_2 are two circles on the torus shown on the picture on p. 105 in the main body of the text, and M and N are integers. The addition in the fundamental group coincides with the usual addition of these expressions.

20. The easiest way to describe the fundamental group of the Riemann surface of a positive genus g (that is, with g holes) is again to realize that we can obtain it by gluing the opposite sides of a polygon – but now with $4g$ vertices. For example, let's glue together the opposite sides of an octagon (the polygon with eight vertices). There are four pairs of opposite sides in this case, and we identify the sides in each pair. The result of this gluing is more difficult to imagine than in the case of the torus, but it is known that we obtain a Riemann surface of genus two (the surface of a Danish pastry).

This can be used to describe the fundamental group of a general Riemann surface similarly to the way we described the fundamental group of a torus. As in the case of a torus, we construct $2g$ elements in the fundamental group of the Riemann surface of genus g by taking the paths along $2g$ consecutive sides of the polygon. (Each of the remaining $2g$ sides will be identified with one of these.) Let us denote them by $a_1, a_2, \ldots, a_{2g}$. They will generate the fundamental group of our Riemann surface, in the sense that any element of this group may be obtained by adding these together, possibly several times. For example, for $g = 2$ we have the following element: $a_3 + 2a_1 + 3a_2 + a_3$. (But note that we cannot rewrite this as $2a_3 + 2a_1 + 3a_2$, because a_3 does not commute with a_2 and a_1, so we cannot move the rightmost a_3 to the left.)

In the same way as in the torus case, by expressing the path connecting two opposite corners of our polygon in two different ways, we obtain a relation between them, generalizing the commutativity relation in the torus case:

$$a_1 + a_2 + \ldots + a_{2g-1} + a_{2g} = a_{2g} + a_{2g-1} + \ldots + a_2 + a_1.$$

This actually turns out to be the only relation between these elements, so we obtain a concise description of the fundamental group: it is generated by $a_1, a_2, \ldots, a_{2g}$, subject to this relation.

21. To explain this more precisely, consider all rational functions on our Riemann surface, in the sense of endnote 13, above. They are analogous to the rational numbers. The relevant Galois group is defined as the group of symmetries of a number field obtained by adjoining solutions of polynomial equations such as $x^2 = 2$ to the rational numbers. Likewise, we can adjoin solutions of polynomial equations to rational functions on a Riemann surface X. It turns out that when we do this, we obtain rational functions on another Riemann surface X', which is a "covering" of X; that is, we have a map $X' \to X$ with finite fibers. In this situation, the Galois group consists of those symmetries of X' which leave all points of X unchanged. In other words, these symmetries act along the fibers of the map $X' \to X$.

Now observe that if we have a closed path on the Riemann surface X, starting and ending at a point P on X, we can take each point of X' in the fiber over P and "follow" it along this path. When we come back, we will in general get a different point in the fiber over P, so we obtain a transformation of this fiber. This is the phenomenon of monodromy, which will be discussed in more detail in Chapter 15. This transformation of the fiber may be traced to an element of the Galois group. Thus, we obtain a link between the fundamental group and the Galois group.

Chapter 10. Being in the Loop

1. The word "special" refers to those orthogonal transformations that preserve orientation – these are precisely the rotations of the sphere. An example of an orthogonal transformation that does not preserve orientation (and hence does not belong to $SO(3)$) is a reflection with respect to one of the coordinate planes. The group $SO(3)$ is closely related to the group $SU(3)$, which we discussed in Chapter 2 in connection with quarks (the special unitary group of the 3-dimensional space). The group $SU(3)$ is defined analogously to $SO(3)$; we replace the *real* 3-dimensional space by the *complex* 3-dimensional space.

2. Yet another way to see that the circle is one-dimensional is to recall that it can be thought of as the set of real solutions of the equation $x^2 + y^2 = 1$, as we discussed in Chapter 9. So the circle is the set of points on the plane constrained by one equation. Hence, its dimension is the dimension of the plane, which is two, minus the number of equations, which is one.

3. This citation appears in the book of Duchamp's notes entitled *À l'Infinitif*, as quoted in Gerald Holton, *Henri Poincaré, Marcel Duchamp and innovation in science and art*, Leonardo, vol. 34, 2001, p. 130.

4. Linda Dalrymple Henderson, *The Fourth Dimension and Non-Euclidean Geometry in Modern Art*, MIT Press, 2013, p. 493.

5. Gerald Holton, ibid., p. 134.

6. Charles Darwin, *Autobiographies*, Penguin Classics, 2002, p. 30.

7. For more details, see for example, Shing-Tung Yau and Steve Nadis, *The Shape of Inner Space*, Basic Books, 2010.

8. It turns out that the dimension of this group is equal to $n(n-1)/2$. In other words, to describe an element of this group we need $n(n-1)/2$ independent coordinates (in the case $n = 3$ we need $3(3-1)/2 = 3$ coordinates, as we have seen in the main text).

9. Mathematically, each loop may be viewed as the image of a particular "map" from the circle to the three-dimensional space, that is, a rule that assigns to each point φ on the circle a point $f(\varphi)$ in the three-dimensional space. We only consider "smooth" maps. Roughly speaking, this means that the loop does not have any sharp angles or corners, and so it looks like the one shown on the picture in the main text.

More generally, a map from a manifold S to a manifold M is a rule that assigns to each point s in S a point in M, called the image of s.

10. See, for example, Brian Greene, *The Elegant Universe*, Vintage Books, 2003.

11. More precisely, a loop in $SO(3)$ is a collection $\{f(\varphi)\}$ of elements of $SO(3)$, parametrized by the angle φ (which is a coordinate on the circle). Given a second loop, which is a collection $\{g(\varphi)\}$, let's compose the two rotations, $f(\varphi) \circ g(\varphi)$ for each φ. Then we get a new collection $\{f(\varphi) \circ g(\varphi)\}$, which is another loop in $SO(3)$. Thus, for each pair of loops in $SO(3)$ we produce a third loop. This is the rule of multiplication in the loop group. The identity element of the loop group is the loop concentrated at the identity of $SO(3)$, that is, $f(\varphi)$ is the identity element of $SO(3)$ for all φ. The inverse loop to the loop $\{f(\varphi)\}$ is the loop $\{f(\varphi)^{-1}\}$. It is easy to check that all axioms of the group hold. Hence, the loop space of $SO(3)$ is indeed a group.

12. To see this, let's consider a simpler example: the loop space of the plane. The plane has two coordinates, x and y. Therefore a loop on the plane is the same as a collection of points on the plane with coordinates $x(\varphi)$ and $y(\varphi)$, one for each angle φ between 0 and 360 degrees. (For example, the formulas $x(\varphi) = \cos(\varphi), y(\varphi) = \sin(\varphi)$ describe a particular loop: the circle of radius 1 centered at the origin.) Therefore to specify such a loop we need to specify an infinite collection of pairs of numbers, $(x(\varphi), y(\varphi))$, one pair for each angle φ. That's why the loop space of the plane is infinite-dimensional. For the same reason, the loop space of any finite-dimensional manifold is also infinite-dimensional.

13. Quoted in R. E. Langer, *René Descartes*, The American Mathematical Monthly, vol. 44, No. 8, October 1937, p. 508.

14. The tangent plane is the closest plane to the sphere among all planes passing through this point. It only touches the sphere at this one point, whereas if we move this plane even slightly (so that it still passes through the same fixed point on the sphere), we obtain a plane that intersects the sphere at more points.

15. By definition, the Lie algebra of a given Lie group is the flat space (such as a line, a plane, and so on) that is the closest to this Lie group among all other flat spaces passing through the point in the Lie group corresponding to the identity.

16. A general circle does not have a special point. But the *circle group* does: it is the identity element of this group, which is a special point of the circle. It must be specified to make a circle into a group.

17. Here is a more precise definition of a vector space:

Once we choose a coordinate system in an n-dimensional flat space, we identify the points of this space with n-tuples of real numbers, $(x_1, x_2, \ldots, x_n)$, the numbers x_i being the coordinates of a point. In particular, there is a special point $(0, 0, \ldots, 0)$, at which all coordinates are equal to 0. This is the origin.

Now fix a point $(x_1, x_2, \ldots, x_n)$ in this space. We define a symmetry of our space, which sends any other point $(z_1, z_2, \ldots, z_n)$ to $(z_1 + x_1, z_2 + x_2, \ldots, z_n + x_n)$. Geometrically, we can think of this symmetry as the shift of our n-dimensional space in the direction of the pointed interval connecting the origin and the point $(x_1, x_2, \ldots, x_n)$. This symmetry is called a *vector*, and it is usually represented by this pointed interval. Let's denote this vector by $\langle x_1, x_2, \ldots, x_n \rangle$. There is a one-to-one correspondence between points of the n-dimensional flat space and vectors. For this reason, the flat space with a fixed coordinate system may be viewed as the space of vectors. Hence, we call it *vector space*.

The advantage of thinking in terms of vectors rather than points is that we have two natural operations on vectors. The first is the operation of addition of vectors, which makes a vector space into a group. As explained in Chapter 2, symmetries can be composed, and therefore they form a group. The composition of the shift symmetries described in the previous paragraph gives us the following rule of addition of vectors:

$$\langle x_1, x_2, \ldots, x_n \rangle + \langle y_1, y_2, \ldots, y_n \rangle = \langle y_1 + x_1, y_2 + x_2, \ldots, y_n + x_n \rangle.$$

The identity element in the group of vectors is the vector $\langle 0, 0, \ldots, 0 \rangle$. The additive inverse of the vector $\langle x_1, x_2, \ldots, x_n \rangle$ is the vector $\langle -x_1, -x_2, \ldots, -x_n \rangle$.

The second is the operation of multiplication of vectors by real numbers. The result of multiplication of a vector $\langle x_1, x_2, \ldots, x_n \rangle$ by a real number k is the vector $\langle kx_1, kx_2, \ldots, kx_n \rangle$.

Thus, a vector space carries two structures: addition, satisfying the properties of a group, and multiplication by numbers. These structures must satisfy natural properties.

Now, any tangent space is a vector space, and therefore any Lie algebra is a vector space.

What is described above is the notion of a vector space over real numbers. Indeed, the coordinates of vectors are real numbers and so we can multiply vectors by real numbers. If we replace real numbers by complex numbers in this description, we obtain the notion of a vector space over complex numbers.

18. The operation on a Lie algebra is usually denoted by square brackets, so if $\vec{a}$ and $\vec{b}$ denote two vectors in a Lie algebra (which is a vector space, as explained in the previous endnote), then the result of this operation on them is denoted by $[\vec{a}, \vec{b}]$. It satisfies the following properties: $[\vec{a}, \vec{b}] = -[\vec{b}, \vec{a}]$, $[\vec{a} + \vec{b}, c] = [\vec{a}, c] + [\vec{b}, c]$, $[k\vec{a}, \vec{b}] = k[\vec{a}, \vec{b}]$ for any number k, and the so-called Jacobi identity:

$$[[\vec{a}, \vec{b}], \vec{c}] + [[\vec{b}, \vec{c}], \vec{a}] + [[\vec{c}, \vec{a}], \vec{b}] = 0.$$

19. The cross-product of two vectors in the three-dimensional space, $\vec{a}$ and $\vec{b}$, is the vector, denoted by $\vec{a} \times \vec{b}$, which is perpendicular to the plane containing $\vec{a}$ and $\vec{b}$, has length equal to the product of the lengths of $\vec{a}$ and $\vec{b}$ and the sine of the angle between them, and such that the triple of vectors $\vec{a}$, $\vec{b}$, and $\vec{a} \times \vec{b}$ has positive orientation (this may be expressed by the so-called right-hand rule).

20. For example, the Lie algebra of the Lie group $SO(3)$ is the three-dimensional vector space. Therefore the Lie algebra of the loop group of $SO(3)$ consists of all loops in this three-dimensional space. The cross-product in the three-dimensional space gives a Lie algebra structure on these loops. Thus, given two loops, we produce a third, even though it's not easy to describe what it is in words.

21. More precisely, a Kac–Moody algebra is an extension of the Lie algebra of a loop group by a one-dimensional space. For more details, see Victor Kac, *Infinite-dimensional Lie Algebras*, Third Edition, Cambridge University Press, 1990.

22. The models with Virasoro algebra symmetry are called conformal field theories, first introduced by the Russian physicists Alexander Belavin, Alexander Polyakov, and Alexander Zamolodchikov in 1984. Their seminal work relied on the results obtained by Feigin and Fuchs, as well as Victor Kac.

23. The most well-known of these are the Wess–Zumino–Witten models. For more details, see Edward Frenkel and David Ben-Zvi, *Vertex Algebras*, Second Edition, American Mathematical Society, 2004.

24. These "quantum fields" have nothing to do with "number fields" or "finite fields" that we discussed in the earlier chapters. This is another example of confusing mathematical terminology, though in other languages there is no confusion: the French, for example, use the word "champs" for quantum fields and "corps" for number fields and finite fields.

Chapter 11. Conquering the Summit

1. Here is a precise construction: suppose we have an element of the loop group of $SO(3)$, which is a collection $\{g(\varphi)\}$ of elements of $SO(3)$ parametrized by the angle φ (the coordinate on the circle). On the other hand, an element of the loop space of the sphere is a collection $\{f(\varphi)\}$ of points of the sphere parametrized by φ. Given such $\{g(\varphi)\}$ and $\{f(\varphi)\}$, we construct another element of the loop space of the sphere as the collection $\{g(\varphi)\big(f(\varphi)\big)\}$. This means that we apply the rotation $g(\varphi)$ to the point $f(\varphi)$ of the sphere, independently for each φ. Thus, we see that each element of the loop group of $SO(3)$ gives rise to a symmetry of the loop space of the sphere.

2. A point of a flag manifold is a collection: a line in a fixed n-dimensional space, a plane that contains this line, the three-dimensional space that contains the plane, and so on, up to an $(n-1)$-dimensional hyperplane containing all of them.

Contrast this with the projective spaces, which I had studied at first: a point of the projective space is just a line in the n-dimensional space, nothing else.

In the simplest case, $n = 2$, our fixed space is two-dimensional, and so the only choice we have is that of a line (there is only one plane, the space itself). Therefore, in this case the flag manifold is the same as the projective space, and it turns out to coincide with the sphere. It is important to note that here we consider lines, planes, and so forth in a complex space (not a real space), and only those that pass through the origin of our fixed n-dimensional space.

The next example is $n = 3$, so we have a three-dimensional space. In this case the projective space consists of all lines in this three-dimensional space, but the flag manifold consists of pairs: a line and a plane containing it (there is only one three-dimensional space). Therefore, in this case there is a difference between the projective space and the flag manifold. We can think of the line as the pole of a flag and the plane as the banner of the flag. Hence the name "flag manifold."

3. Boris Feigin and Edward Frenkel, *A family of representations of affine Lie algebras*, Russian Mathematical Surveys, vol. 43, No. 5, 1988, pp. 221–222.

Chapter 12. Tree of Knowledge

1. Mark Saul, *Kerosinka: An episode in the history of Soviet mathematics*, Notices of the American Mathematical Society, vol. 46, November 1999, pp. 1217–1220.

2. I learned later that Gelfand, who collaborated with heart doctors (for the same reason as Yakov Isaevich collaborated with urologists), also successfully used this approach to medical research.

Chapter 14. Tying the Sheaves of Wisdom

1. A precise definition of a vector space was given in endnote 17 to Chapter 10.

2. In the case of the category of vector spaces, morphisms from a vector space V_1 to a vector space V_2 are the so-called linear transformations from V_1 to V_2. These are maps f from V_1 to V_2 such that $f(\vec{a} + \vec{b}) = f(\vec{a}) + f(\vec{b})$ for any two vectors $\vec{a}$ and $\vec{b}$ in V_1, and $f(k \cdot \vec{a}) = k \cdot f(\vec{a})$ for any vector $\vec{a}$ in V_1 and number k. In particular, the morphisms from a given vector space V to itself are the linear transformations from V to itself. The symmetry group of V consists of those morphisms that have an inverse.

3. See, for example, Benjamin C. Pierce, *Basic Category Theory for Computer Scientists*, MIT Press, 1991.

Joseph Goguen, *A categorical manifesto*, Mathematical Structures in Computer Science, vol. 1, 1991, 49–67.

Steve Awodey, *Category Theory*, Oxford University Press, 2010.

4. See, for example, http://www.haskell.org/haskellwiki/Category_theory and references therein.

5. See, for example, Masaki Kashiwara and Pierre Schapira, *Sheaves on Manifolds*, Springer-Verlag, 2010.

6. This surprising property of arithmetic modulo primes has a simple explanation if we view it from the point of view of group theory. Consider the non-zero elements of the finite field: $1, 2, ..., p-1$. They form a group with respect to multiplication. Indeed, the identity element with respect to multiplication is the number 1: if we multiply any element a by 1, we get back a. And each element has an inverse, as explained in endnote 8 to Chapter 8: for any a in $\{1, 2, ..., p-1\}$, there is an element b such that $a \cdot b = 1$ modulo p.

This group has $p-1$ elements. There is a general fact that holds for *any* finite group G with N elements: the Nth power of each element a of this group is equal to the identity element (which we will denote by 1),

$$a^N = 1.$$

To prove this, consider the following elements in the group G: $1, a, a^2, \ldots$ Because the group G is finite, these elements cannot all be distinct. There have to be repetitions. Let k be the smallest natural number such that a^k is equal to 1 or a^j for some $j = 1, \ldots, k-1$. Suppose that the latter is the case. Let a^{-1} denote the inverse of a, so that $a \cdot a^{-1} = 1$ and take its jth power $(a^{-1})^j$. Multiply both sides of the equation $a^k = a^j$ with $(a^{-1})^j$ on the right. Each time we encounter $a \cdot a^{-1}$ we replace it by 1. Multiplying by 1 does not change the result, so we can always remove 1 from the product. We see then that each a^{-1} will cancel one of the a's. Hence, the left-hand side will be equal to a^{k-j}, and the right-hand side will be equal to 1. We obtain that $a^{k-j} = 1$. But $k-j$ is smaller than k, and this contradicts our choice of k. Therefore, the first repetition on our list will necessarily have the form $a^k = 1$, so that the elements $1, a, a^2, \ldots, a^{k-1}$ are all distinct. This means that they form a group of k elements: $\{1, a, a^2, \ldots, a^{k-1}\}$. It is a subgroup of our original group G of N elements, in the sense that it is a subset of elements of G such that the result of multiplication of any two elements of this subset is again an element of the subset, this subset contains the identity element of G, and this subset contains the inverse of each of its elements.

Now, it is known that the number of elements of any subgroup always divides the number of elements of the group. This statement is called the Lagrange theorem. I'll leave it for you to prove (or you may just Google it).

Applying the Lagrange theorem to the subgroup $\{1, a, a^2, \ldots, a^{k-1}\}$, which has k elements, we find that k must divide N, the number of elements of the group G. Thus, $N = km$ for some natural number m. But since $a^k = 1$, we obtain that

$$a^N = (a^k) \cdot (a^k) \cdot \ldots \cdot (a^k) = 1 \cdot 1 \cdot \ldots \cdot 1 = 1,$$

which is what we wanted to prove.

Let's go back to the group $\{1, 2, \ldots, p-1\}$ with respect to multiplication. It has $p-1$ elements. This is our group G, so our N is equal to $p-1$. Applying the general result in this case, we find that $a^{p-1} = 1$ modulo p for all a in $\{1, 2, \ldots, p-1\}$. But then

$$a^p = a \cdot a^{p-1} = a \cdot 1 = a \quad \text{modulo} \quad p.$$

It is easy to see that the last formula actually holds for any integer a, if we stipulate that

$$x = y \quad \text{modulo} \quad p$$

whenever $x - y = rp$ for some integer r.

This is the statement of Fermat's little theorem. Fermat first stated it in a letter to his friend. "I would send you a proof," he wrote, "but I am afraid it's too long."

7. Up to now, we have considered the arithmetic modulo a prime number p. However, it turns out that there is a statement analogous to Fermat's little theorem in the arithmetic modulo any natural number n. To explain what it is, I need to recall the Euler function ϕ, which we discussed in conjunction with braid groups in Chapter 6. (In my braid group project, I had found that the Betti numbers of braid groups are expressed in terms of this function.) I recall that $\phi(n)$ is the number of natural numbers between 1 and $n-1$ that are relatively prime with n; that is, do not have common divisors with n (other than 1). For instance, if n is a prime, then all numbers between 1 and $n-1$ are relatively prime to n, and so $\phi(n) = n - 1$.

Now, the analogue of the formula $a^{p-1} = 1$ modulo p that we proved in the previous endnote is the formula

$$a^{\phi(n)} = 1 \quad \text{modulo} \quad n.$$

It holds for any natural number n and any natural number a that is relatively prime to n. It is proved in exactly the same way as before: we take the set of all natural numbers between 1 and $n-1$ that are relatively prime to n. There are $\phi(n)$ of them. It is easy to see that they form a group with respect to the operation of multiplication. Hence, by the Lagrange theorem, for any element of this group, its $\phi(n)$th power is equal to the identity element.

Consider, for example, the case that n is the product of two prime numbers. That is, $n = pq$, where p and q are two distinct prime numbers. In this case, the numbers that are not relatively prime to n are either divisible by p or by q. The former have the form pi, where $i = 1, \ldots, q-1$ (there are $q-1$ of those), and the latter have the form qj, where $j = 1, \ldots, p-1$ (there are $p-1$ of those). Hence we find that

$$\phi(n) = (n-1) - (q-1) - (p-1) = (p-1)(q-1).$$

Therefore we have

$$a^{(p-1)(q-1)} = 1 \quad \text{modulo} \quad pq$$

for any number a that is not divisible by p and q. And it is easy to see that the formula

$$a^{1+m(p-1)(q-1)} = a \quad \text{modulo} \quad pq$$

is true for any natural number a and any integer m.

This equation is the basis of one of the most widely used encryption algorithms, called RSA algorithm (after Ron Rivest, Adi Shamir, and Leonard Adleman, who described it in 1977). The idea is that we pick two primes p and q (there are various algorithms for generating them) and let n be the product pq. Number n is made public, but the primes p and q are not. Next, we pick a number e that is relatively prime to $(p-1)(q-1)$. This number is also made public.

The encryption process converts any number a (such as a credit card number) to a^e modulo n:

$$a \quad \mapsto \quad b = a^e \quad \text{modulo} \quad n.$$

It turns put that there is an efficient way to reconstruct a from a^e. Namely, we find a number d between 1 and $(p-1)(q-1)$ such that

$$de = 1 \quad \text{modulo} \quad (p-1)(q-1).$$

In other words,

$$de = 1 + m(p-1)(q-1)$$

for some natural number m. Then

$$a^{de} \quad \text{modulo} \quad n \quad = \quad a^{1+m(p-1)(q-1)} \quad \text{modulo} \quad n$$
$$= \quad a \quad \text{modulo} \quad n$$

according to the formula above.

Therefore, given $b = a^e$, we can recover the original number a as follows:

$$b \quad \mapsto \quad b^d \quad \text{modulo} \quad n.$$

Let's summarize: we make the numbers n and e public, but keep d secret. The encryption is given by the formula

$$a \quad \mapsto \quad b = a^e \quad \text{modulo} \quad n.$$

Anyone can do it because e and n are publicly available.

The decryption is given by the formula

$$b \quad \mapsto \quad b^d \quad \text{modulo} \quad n.$$

Applied to a^e, it gives us back the original number a. But only those who know d can do this.

The reason why this is a good encryption scheme is that in order to find d, which enables one to reconstruct the numbers being encoded, we must know the value of $(p-1)(q-1)$. But for this we need to know what p and q are, which are the two prime divisors of n. Those are kept secret. For n sufficiently large, using known methods of prime factorization, it may take many months, even on a network of powerful computers, to find p and q. For example, in 2009 a group of researchers using hundreds of parallel computers were able to factor into primes a 232-digit number; it took two years (see http://eprint.iacr.org/2010/006.pdf). But if one could come up with a more efficient way to "factor" natural numbers into primes (for example, by using a quantum computer), then one would have a tool for breaking this encryption scheme. That's why a lot of research is directed toward factoring numbers into primes.

8. In the case of the rational numbers, we saw that the equations of the form $x^2 = 2$ may not have solutions among rational numbers, and in that case we could create a new numerical system by adding these solutions, such as $\sqrt{2}$ and $-\sqrt{2}$. Then we saw that flipping $\sqrt{2}$ and $-\sqrt{2}$ was a symmetry of this new numerical system.

Likewise, we can consider polynomial equations in the variable x, such as $x^2 = 2$ or $x^3 - x = 1$, as equations in the finite field $\{0, 1, 2, ..., p - 1\}$. We may then ask whether this equation can be solved for x inside this finite field. If it does not have a solution, then we can adjoin the solutions to the finite field, in the same way in which we adjoined $\sqrt{2}$ and $-\sqrt{2}$ to the rational numbers. This way we create new finite fields.

For instance, if $p = 7$, the equation $x^2 = 2$ has two solutions, 3 and 4, because

$$3^2 = 9 = 2 \quad \text{modulo} \quad 7, \qquad 4^2 = 16 = 2 \quad \text{modulo} \quad 7.$$

Note that 4 is -3 with respect to the arithmetic modulo 7 because $3 + 4 = 0$ modulo 7. So these two solutions of the equation $x^2 = 2$ are the negatives of each other, just like $\sqrt{2}$ and $-\sqrt{2}$ are the negatives of each other. This is not surprising: the two solutions of the equation $x^2 = 2$ will always be the negatives of each other because if $a^2 = 2$, then $(-a)^2 = (-1)^2 a^2 = 2$ as well. This means that if $p \neq 2$, there will always be two elements of the finite field that square to the same number, and they will be the negatives of each other (if $p \neq 2$, then p is necessarily odd, and then $-a$ cannot be equal to a, for otherwise p would be equal to $2a$). Therefore, only half of the non-zero elements of the finite field $\{1, 2, ..., p - 1\}$ are squares.

(The celebrated Gauss reciprocity law describes which numbers n are squares in the arithmetic modulo p and which are not. This is beyond the scope of this book, but let's just say that the answer only depends on the value of p modulo $4n$. So, for example, we already know that

$n = 2$ is a square modulo $p = 7$. In this case $4n = 8$. Hence, it will also be a square modulo any prime p which is equal to 7 modulo 8, no matter how large. This is a stunning result!)

If $p = 5$, then $1^2 = 1$, $2^2 = 4$, $3^2 = 4$, and $4^2 = 1$ modulo 5. So 1 and 4 are squares modulo 5, but 2 and 3 are not. In particular, we see that there are no solutions to the equation $x^2 = 2$ in the finite field $\{0, 1, 2, 3, 4\}$, just as it was in the case of rational numbers. Hence, we can create a new numerical system extending the finite field $\{0, 1, 2, 3, 4\}$ by adjoining the solutions of $x^2 = 2$. Let us denote them again by $\sqrt{2}$ and $-\sqrt{2}$ (but we have to remember that these are not the same numbers that we adjoined to the rational numbers before).

We obtain a new finite field which consists of numbers of the form

$$a + b\sqrt{2},$$

where a and b are in $\{0, 1, 2, 3, 4\}$. Because we have two parameters a and b taking values 0, 1, 2, 3, or 4, we find that this new numerical system has $5 \cdot 5 = 25$ elements. More generally, any finite extension of the field $\{0, 1, ..., p-1\}$ has p^m elements for some natural number m.

Now suppose that we adjoin all solutions of all polynomial equations in one variable to the finite field $\{0, 1, 2, ..., p-1\}$. Then we obtain a new numerical system called the *algebraic closure* of the finite field. The original finite field has p elements. It turns out that its algebraic closure has infinitely many elements. Our next question is what is the Galois group of this algebraic closure. These are the symmetries of this algebraic closure, which preserve the operations of addition and multiplication and send the elements of the original field of p elements to themselves.

If we start with the field of rational numbers and take its algebraic closure, then the corresponding Galois group is very complicated. In fact, the Langlands Program was created in part to describe this Galois group and its representations in terms of harmonic analysis.

In contrast, the Galois group of the algebraic closure of the finite field $\{0, 1, 2, ..., p-1\}$ turns out to be quite simple. Namely, we already know one of these symmetries: the Frobenius, which is the operation of raising to the pth power: $a \mapsto a^p$. According to Fermat's little theorem, the Frobenius preserves all elements of the original finite field of p elements. It also preserves addition and multiplication in the algebraic closure:

$$(a + b)^p = a^p + b^p, \qquad (ab)^p = a^p b^p.$$

Hence, the Frobenius belongs to the Galois group of the algebraic closure of the finite field.

Let us denote the Frobenius by F. Clearly, any integer power F^n of the Frobenius is also an element of the Galois group. For example, F^2 is the operation of raising to the p^2th power, $a \mapsto a^{p^2} = (a^p)^p$. The symmetries F^n, where n runs over all integers, form a subgroup of the Galois group which is called the Weil group, in honor of André Weil. The Galois group itself is what's called a completion of the Weil group; in addition to the integer powers of F, it also has as elements certain limits of F^n as n goes to ∞. But in the appropriate sense, the Frobenius generates the Galois group.

Here is an example of how the Frobenius acts on elements of the algebraic closure of a finite field. Consider the case $p = 5$ and the elements of the algebraic closure of the above form

$$a + b\sqrt{2},$$

where a and b are 0, 1, 2, 3, or 4. This numerical system has a symmetry exchanging $\sqrt{2}$ and $-\sqrt{2}$:

$$a + b\sqrt{2} \quad \mapsto \quad a - b\sqrt{2},$$

in parallel with what happens when we adjoin $\sqrt{2}$ to the rational numbers. What is surprising (and what has no analogue in the case of rational numbers) is that this flip symmetry is in fact equal to the Frobenius. Indeed, applying the Frobenius to $\sqrt{2}$ means raising it to the 5th power, and we find that

$$(\sqrt{2})^5 = (\sqrt{2})^2 \cdot (\sqrt{2})^2 \cdot \sqrt{2} = 2 \cdot 2 \cdot \sqrt{2} = 4 \cdot \sqrt{2} = -\sqrt{2},$$

because $4 = -1$ modulo 5. It follows that for $p = 5$ the Frobenius sends $a + b\sqrt{2}$ to $a - b\sqrt{2}$. The same is true for any prime p such that the equation $x^2 = 2$ has no solutions in the finite field $\{0, 1, 2, ..., p - 1\}$.

9. A symmetry of an n-dimensional vector space, more properly called linear transformation (see endnote 2), may be represented by a matrix, that is a square array of numbers a_{ij}, where i and j run from 1 to n, where n is the dimension of the vector space. Then the trace is the sum of the diagonal elements of this matrix, that is, of all a_{ii} with i running from 1 to n.

10. In the present context, going back would mean finding, for a given function f, a sheaf such that for each point s in our manifold, the trace of the Frobenius on the fiber at s is equal to value of f at s. Any given number can be realized as the trace of a symmetry of a vector space. What is difficult is to combine these vector spaces into a coherent collection satisfying the properties of a sheaf.

Chapter 15. A Delicate Dance

1. A representation of the Galois group in a group H is a rule that assigns to each element of the Galois group an element of H. It should satisfy the condition that if a, b are two elements of the Galois group and $f(a), f(b)$ are the elements of H assigned to them, then to the product ab in the Galois group should be assigned the product $f(a)f(b)$ in H. A more proper name for this is a *homomorphism* from the Galois group to H.

2. To make this more precise, recall the notion of an n-dimensional vector space from endnote 17 to Chapter 10. As we discussed in Chapter 2, an n-dimensional representation of a given group is a rule that assigns a symmetry S_g of an n-dimensional vector space to each element g of this group. This rule must satisfy the following property: for any two elements of the group, g and h, and their product gh in the group, the symmetry S_{gh} is equal to the composition of S_g and S_h. It is also required that for each element g, we have $S_g(\vec{a} + \vec{b}) = S_g(\vec{a}) + S_g(\vec{b})$ and $S_g(k \cdot \vec{a}) = k \cdot S_a(\vec{a})$ for any vectors $\vec{a}, \vec{b}$ and a number k. (Such symmetries are called linear transformations; see endnote 2 to Chapter 14.)

The group of all invertible linear transformations of an n-dimensional vector space is called the general linear group. It is denoted by $GL(n)$. Thus, according to the definition in the previous paragraph, an n-dimensional representation of a given group Γ is the same as a representation of Γ in $GL(n)$ (or, a homomorphism from Γ to $GL(n)$, see endnote 1).

For example, in Chapter 10 we talked about a three-dimensional representation of the group $SO(3)$. Each element of the group $SO(3)$ is a rotation of a sphere, to which we assign the corresponding rotation of the three-dimensional vector space containing the sphere (it turns out to be a linear transformation). This gives us a representation of $SO(3)$ in $GL(3)$ (or equivalently, a homomorphism from $SO(3)$ to $GL(3)$). Intuitively, we can think about a rotation as "acting" on the three-dimensional vector space, rotating each vector in this space to another vector.

On one side of the Langlands relation (also known as the Langlands correspondence), we consider n-dimensional representations of the Galois group. On the other side, we have automorphic functions that can be used to build the so-called automorphic representations of another group $GL(n)$ of symmetries of the n-dimensional vector space, though not over the real numbers but over the so-called adèles. I will not attempt to explain what these are, but the following diagram shows schematically what the Langlands relation should look like:

n-dimensional representations of the Galois group	$\longleftrightarrow$	automorphic representations of the group $GL(n)$

For example, two-dimensional representations of the Galois group are related to the automorphic representations of the group $GL(2)$, which can be constructed from the modular forms discussed in Chapter 9.

A generalization of this relation is obtained by replacing the group $GL(n)$ by a more general Lie group. Then, on the right-hand side of the relation we have automorphic representations of G, rather than $GL(n)$. On the left-hand side, we have representations of the Galois group in the Langlands dual group ^LG, instead of $GL(n)$ (or equivalently, homomorphisms from the Galois group to ^LG). For more details, see, for example, my survey article: Edward Frenkel, *Lectures on the Langlands Program and conformal field theory*, in *Frontiers in Number Theory, Physics and Geometry II*, eds. P. Cartier, e.a., pp. 387–536, Springer-Verlag, 2007, available online at

http://arxiv.org/pdf/hep-th/0512172.pdf

3. See the video at http://www.youtube.com/watch?v=CYBqIRM8GiY

4. This dance is called Binasuan. See, for example, this video:

http://www.youtube.com/watch?v=N2TOOz_eaTY

5. For the construction of this path and the explanation why if we traverse this path twice, we obtain a trivial path, see, for example, Louis H. Kaufmann, *Knots and Physics*, Third Edition, pp. 419–420, World Scientific, 2001.

6. In other words, the fundamental group of $SO(3)$ consists of two elements: one is the identity and the other is this path, whose square is the identity.

7. The mathematical name for this group is $SU(2)$. It consists of the "special unitary" transformations of the two-dimensional complex vector space. This group is a cousin of the group $SU(3)$, discussed in Chapter 2 in connection with quarks, which consists of special unitary transformations of the three-dimensional complex vector space.

8. More precisely, the lifting of the closed path we have constructed (corresponding to the first full turn of the cup) from the group $SO(3)$ to its double cover, the group $SU(2)$, will be a path that starts and ends at different points of $SU(2)$ (both of which project onto the same point in $SO(3)$), so it's not a closed path in $SU(2)$.

9. In general, this relationship is more subtle, but to simplify matters, in this book we will assume that the dual of the dual group is the group itself.

10. A principal G-bundle (or, G-bundle) on a Riemann surface is a fibration over the Riemann surface such that all fibers are copies of the "complexification" of the group G (it is defined by replacing, in the definition of the group, real numbers by complex numbers). The points of the moduli space (more properly called stack) of G-bundles on X are equivalence classes of G-bundles on X.

In order to simplify the exposition, in this book we do not make a distinction between a Lie group and its complexification.

11. In the fundamental group, we identify any two closed paths that could be deformed into one another. Since any closed path on the plane that does not go around the removed point can be contracted to a point, the non-trivial elements of the fundamental group are those closed paths that go around this point (those cannot be contracted – the point we have removed from the plane is an obstacle to doing this).

It is easy to see that any two closed paths with the same winding number can be deformed into one another. So the fundamental group of the plane without one point is nothing but the group of integers. Note that this discussion is reminiscent of our discussion in Chapter 5 of the braid group with two threads, which we also found to be the same as the group of integers. This is not a coincidence because the space of pairs of distinct points on the plane is topologically equivalent to the plane with a point removed.

12. The reason that the monodromy takes values in the circle group is due to the famous Euler's formula

$$e^{\theta\sqrt{-1}} = \cos(\theta) + \sin(\theta)\sqrt{-1}.$$

In other words, the complex number $e^{\theta\sqrt{-1}}$ is represented by the point on the unit circle corresponding to the angle θ, as measured in radians. Recall that 2π radians is the same as 360 degrees (this corresponds to the full rotation of the circle). Therefore, the angle θ measured in radians is the angle $360 \cdot \theta/2\pi$ degrees.

A special case of this formula, for $\theta = \pi$, is

$$e^{\pi\sqrt{-1}} = -1,$$

which Richard Feynman called "one of the most remarkable, almost astounding, formulas in all of mathematics." It played a prominent role in the Yoko Ogawa's novel *The Housekeeper and the Professor*, Picador, 2009. Another special case, no less important, is $e^{2\pi\sqrt{-1}} = 1$.

This means that the unit circle in the complex plane with the coordinate t, on which the solution of our differential equation is defined, consists of all points of the form $t = e^{\theta\sqrt{-1}}$, where θ is between 0 and 2π. As we move along the unit circle counterclockwise, we are evaluating our solution $x(t) = t^n$ at these points $t = e^{\theta\sqrt{-1}}$, letting the angle θ increase from 0 to 2π (in radians). Making full circle means setting θ equal 2π. Therefore, to obtain the corresponding value of our solution, we need to substitute $t = e^{2\pi\sqrt{-1}}$ into t^n. The result is $e^{2\pi n\sqrt{-1}}$. But the original value of the solution is obtained by substituting $t = 1$ into t^n, which is 1. Thus, we find that when we traverse the closed path going counterclockwise along the unit circle, our solution gets multiplied by $e^{2\pi n\sqrt{-1}}$. And that's the monodromy along this path.

This monodromy $e^{2\pi n\sqrt{-1}}$ is a complex number that can be represented by a point on the unit circle on *another* complex plane. That point corresponds to the angle $2\pi n$ radians, or $360n$ degrees, which is what we wanted to show. In fact, multiplying any complex number z by $e^{2\pi n\sqrt{-1}}$ amounts, geometrically, to rotating the point on the plane corresponding to z by $360n$ degrees. If n is an integer, then $e^{2\pi n\sqrt{-1}} = 1$, so no monodromy occurs, but if n is not an integer, we get a non-trivial monodromy.

To avoid confusion, I want to stress that we have two different complex planes here: one is the complex plane on which our solution is defined – the "t-plane." The other is the plane on which we represent the monodromy. It has nothing to do with the t-plane.

To recap, we have interpreted the monodromy of the solution along a closed path with the winding number +1 on the t-plane as a point of another unit circle. Similarly, if the winding number of the path is w, then the monodromy along this path is $e^{2\pi wn\sqrt{-1}}$, which amounts to the rotation by $2\pi nw$ radians, or $360wn$ degrees. Thus, the monodromy gives rise to a representation of the fundamental group in the circle group. Under this representation, the path on the t-plane without a point, whose winding number is w, goes to the rotation by $360wn$ degrees.

13. Note that it is important that we have removed one point, the origin, from the plane. Otherwise, any path on the plane could be collapsed, and the fundamental group would be trivial. Hence no monodromy would be possible. We are forced to remove this point because our solution, t^n, is not defined at the origin if n is not a natural number or 0 (in that case there is no monodromy).

14. More precisely, not all representations of the fundamental group in $^L G$ can be obtained from opers, and in this diagram we restrict ourselves to those that can. For other representations, the question is still open.

15. Edward Frenkel, *Langlands Correspondence for Loop Groups*, Cambridge University Press, 2007. Online version is available at http://math.berkeley.edu/~frenkel

Chapter 16. Quantum Duality

1. You may be wondering what happened between 1991 and 2003. Well, in this book my main goal is to tell you about the aspects of the Langlands Program that I find most interesting and how the discoveries in this area were made, in which I was fortunate to participate. I am not trying to recount the complete story of my life to date. But, if you are curious, during those years I brought my family from Russia to the U.S., moved out West to Berkeley, California, fell in and out of love, got married and divorced, brought up several Ph.D. students, traveled and lectured around the world, published a book and dozens of research papers. I kept trying to uncover the mysteries of the Langlands Program in different domains: from geometry to integrable systems, from quantum groups to physics. I'll save the details of this part of my journey for another book.

2. See http://www.darpa.mil/Our_Work

3. G.H. Hardy, *A Mathematician's Apology*, Cambridge University Press, 2009, p. 135.

4. R.R. Wilson's Congressional Testimony, April 17, 1969, quoted from http://history.fnal.gov/testimony.html

5. Maxwell's equations in vacuum have the form

$$\nabla \cdot \vec{E} = 0 \qquad\qquad \nabla \cdot \vec{B} = 0$$

$$\nabla \times \vec{E} = -\frac{\partial \vec{B}}{\partial t} \qquad\qquad \nabla \times \vec{B} = \frac{\partial \vec{E}}{\partial t}$$

where $\vec{E}$ denotes the electric field and $\vec{B}$ denotes the magnetic field (to simplify the formulas, we are choosing a system of units in which the speed of light is equal to 1). It is clear that if we send

$$\vec{E} \mapsto \vec{B}, \qquad\qquad \vec{B} \mapsto -\vec{E}$$

the equations of the left-hand side will become the equations on the right-hand side, and vice versa. Thus, each individual equation changes, but the system of equations does not.

6. See Dayna Mason's Flickr page: http://www.flickr.com/photos/daynoir

7. This gauge group $SU(3)$ should not be confused with another group $SU(3)$ discussed in Chapter 2, which was used by Gell-Mann and others to classify elementary particles (it is called the "flavor group"). Gauge group $SU(3)$ has to do with a characteristic of quarks called "color." It turns out that each quark can have three different colors, and gauge group $SU(3)$ is responsible for changing those colors. Because of that, the gauge theory describing the interaction of quarks is called quantum chromodynamics. David Gross, David Politzer, and Frank Wilczek were awarded a Nobel Prize for their stunning discovery of the so-called asympthotic freedom in quantum chromodynamics (and other non-abelian gauge theories), which helped explain quarks' mysterious behavior.

8. D.Z. Zhang, *C.N. Yang and contemporary mathematics*, Mathematical Intelligencer, vol. 15, No. 4, 1993, pp. 13–21.

9. Albert Einstein, *Geometry and Experience*, Address to the Prussian Academy of Sciences in Berlin, January 27, 1921. Translated in G. Jeffrey and W. Perrett, *Geometry and Experience in Sidelights on Relativity*, Methuen, 1923.

10. Eugene Wigner, *The unreasonable effectiveness of mathematics in the natural sciences*, Communications on Pure and Applied Mathematics, Vol. 13, 1960, pp. 1–14.

11. C. Montonen and D. Olive, *Magnetic monopoles as gauge particles?* Physics Letters B, vol. 72, 1977, pp. 117–120.

12. P. Goddard, J. Nuyts, and D. Olive, *Gauge theories and magnetic charge*, Nuclear Physics B, vol. 125, 1977, pp. 1–28.

13. S_e is the set of complex one-dimensional representations of the maximal torus of G, and S_m is the fundamental group of the maximal torus of G. If G is the circle group, then its maximal torus is the circle group itself, and each of these two sets is in one-to-one correspondence with the set of integers.

Chapter 17. Uncovering Hidden Connections

1. The space $M(X,G)$ may be described in several ways; for example, as the space of solutions of a system of differential equations on X first studied by Hitchin (see the article in endnote 19 below for more details). A description that will be useful to us in this chapter is that $M(X,G)$ is the moduli space of representations of the fundamental group of the Riemann surface S in the complexification of the group G (see endnote 10 to Chapter 15). This means that such a representation is assigned to each point of $M(X,G)$.

2. See the video of Hitchin's lecture at the Fields Institute:

http://www.fields.utoronto.ca/video-archive/2012/10/108-690

3. Here I am referring to the recent work of Ngô Bao Châu on the proof of the "fundamental lemma" of the Langlands Program. See, for example, this survey article: David Nadler, *The geometric nature of the fundamental lemma*, Bulletin of the American Mathematical Society, vol. 49, 2012, pp. 1–50.

4. Recall that in sigma model, everything is computed by summing over all maps from a fixed Riemann surface Σ to the target manifold S. In string theory, we make one more step: in addition to summing over all maps from a fixed Σ to S, as we normally do in the sigma model, we also sum up further over all possible Riemann surfaces Σ (the target manifold S remains fixed throughout – this is our space-time). In particular, we sum over the Riemann surfaces of arbitrary genus.

5. For more on superstring theory, see Brian Greene, *The Elegant Universe*, Vintage Books, 2003; *The Fabric of the Cosmos: Space, Time, and the Texture of Reality*, Vintage Books, 2005.

6. For more on Calabi-Yau manifolds and their role in superstring theory, see Shing-Tung Yau and Steve Nadis, *The Shape of Inner Space*, Basic Books, 2010, Chapter 6.

7. A torus also has two continuous parameters: essentially, the radii R_1 and R_2 that we discuss in this chapter, but we will ignore them for the purpose of this discussion.

8. One resolution that has been actively discussed recently is the idea that each of these manifolds gives rise to its own universe with its own physical laws. This is then coupled with a version of the anthropic principle: our universe is selected among them by the fact that physical laws in it allow for intelligent life (so that the question "why is our universe like this?" could be asked). However, this idea, dubbed "string theory landscape" or "multiverse," has been met with a lot of skepticism on both scientific and philosophical grounds.

9. Many interesting properties of quantum field theories in various dimensions have been discovered or elucidated by connecting these theories to superstring theory, using dimensional reduction or studying branes. In a sense, superstring theory has been used as a factory for producing and analyzing quantum field theories (mostly, supersymmetric). For example, this way one obtains a beautiful interpretation of the electromagnetic duality of four-dimensional supersymmetric gauge theories. So, even though we don't know yet whether superstring theory can describe the physics of our universe (and still don't fully understand what superstring theory is), it has already produced many powerful insights into quantum field theory. It has also led to numerous advances in mathematics.

10. The dimension of the Hitchin moduli space $M(X,G)$ is equal to the product of the dimension of the group G (which is the same as the dimension of $^L G$) and $(g-1)$, where g denotes the genus of the Riemann surface X.

11. For more on branes, see Lisa Randall, *Warped Passages: Unraveling the Mysteries of the Universe's Hidden Dimensions*, Harper Perennial, 2006; especially, Chapter IV.

12. More precisely, the A-branes on $M(X,G)$ are objects of a category, the concept we discussed in Chapter 14. The B-branes on $M(X,{}^L G)$ are objects of another category. The statement of homological mirror symmetry is that these two categories are equivalent to each other.

13. Anton Kapustin and Edward Witten, *Electric-magnetic duality and the geometric Langlands Program*, Communications in Number Theory and Physics, vol. 1, 2007, pp. 1–236.

14. For more on the T–duality, see Chapter 7 of the book by Yau and Nadis referenced in endnote 6.

15. For more on the SYZ conjecture, see Chapter 7 of the book by Yau and Nadis referenced in endnote 6.

16. More precisely, each fiber is the product of n circles, where n is an even natural number, so it is an n-dimensional analogue of a two-dimensional torus. Note also that the dimension of the base of the Hitchin fibration and the dimension of each toric fiber will always be equal to each other.

17. In Chapter 15, we discussed a different construction, in which automorphic sheaves were obtained from representations of Kac–Moody algebras. It is expected that the two constructions are related, but as of the time of writing, this relation was still unknown.

18. Edward Frenkel and Edward Witten, *Geometric endoscopy and mirror symmetry*, Communications in Number Theory and Physics, vol. 2, 2008, pp. 113–283, available online at http://arxiv.org/pdf/0710.5939.pdf

19. Edward Frenkel, *Gauge theory and Langlands duality*, Astérisque, vol. 332, 2010, pp. 369–403, available online at http://arxiv.org/pdf/0906.2747.pdf

20. Henry David Thoreau, *A Week on the Concord and Merrimack Rivers*, Penguin Classics, 1998, p. 291.

Chapter 18. Searching for the Formula of Love

1. C.P. Snow, *The Two Cultures*, Cambridge University Press, 1998.

2. Thomas Farber and Edward Frenkel, *The Two-Body Problem*, Andrea Young Arts, 2012. See http://thetwobodyproblem.com/ for more details.

3. Michael Harris, *Further investigations of the mind–body problem*, a chapter from an upcoiming book, available online at http://www.math.jussieu.fr/~harris/MindBody.pdf

4. Henry David Thoreau, *A Week on the Concord and Merrimack Rivers*, Penguin Classics, 1998, p. 291.

5. E.T. Bell, *Men of Mathematics*, Touchstone, 1986, p. 16.

6. Robert Langlands, *Is there beauty in mathematical theories?*, in *The Many Faces of Beauty*, ed. Vittorio Hösle, University of Notre Dame Press, 2013, available online at

http://publications.ias.edu/sites/default/files/ND.pdf

7. Yuri I. Manin, *Mathematics as Metaphor: Selected Essays*, American Mathematical Society, 2007, p. 4.

8. Philosophers have debated the ontology of mathematics for centuries. The point of view that I advocate in this book is usually referred to as mathematical Platonism. Note however that there are different kinds of Platonism, and there are also other philosophical interpretations of mathematics. See, for example, Mark Balaguer, *Mathematical Platonism*, in *Proof and Other Dilemmas: Mathematics and Philosophy*, Bonnie Gold and Roger Simons (eds.), Mathematics Association of America, pp. 179–204, and references therein.

9. Roger Penrose, *The Road to Reality*, Vintage Books, 2004, p. 15.

10. Ibid., pp. 13–14.

11. Kurt Gödel, *Collected Works*, volume III, Oxford University Press, 1995, p. 320.

12. Ibid., p. 323.

13. Roger Penrose, *Shadows of the Mind*, Oxford University Press, 1994, Section 8.47.

14. In the landmark *Gottschalk v. Benson* decision, 409 U.S. 63 (1972), the U.S. Supreme Court stated (quoting earlier cases before the court): "a scientific truth, or the mathematical expression of it, is not a patentable invention.... A principle, in the abstract, is a fundamental truth; an original cause; a motive; these cannot be patented, as no one can claim in either of them an exclusive right.... He who discovers a hitherto unknown phenomenon of nature has no claim to a monopoly of it which the law recognizes."

15. Edward Frenkel, Andrey Losev, and Nikita Nekrasov, *Instantons beyond topological theory I*, Journal of the Institute of Mathematics of Jussieu, vol. 10, 2011, 463–565; there is a footnote in the article explaining that formula (5.7) played the role of "formula of love" in *Rites of Love and Math*.

16. We consider the supersymmetric quantum mechanical model on the sphere (denoted here by $\mathbb{P}^1$) and the correlation function of two observables, denoted by F and ω. This correlation function is defined in our theory as the integral appearing on the left-hand side of the formula. However, our theory also predicts a different expression for it: a sum over the "intermediate states" appearing on the right-hand side. Consistency of our theory requires that the two sides be equal to each other. And indeed they are; that's what our formula says.

17. *Le Monde Magazine*, April 10, 2010, p. 64.

18. Laura Spinney, *Erotic equations: Love meets mathematics on film*, New Scientist, April 13, 2010, available online at http://ritesofloveandmath.com

19. Hervé Lehning, *La dualité entre l'amour et les maths*, Tangente Sup, vol. 55, May–June 2010, pp. 6–8, available online at http://ritesofloveandmath.com

20. We used the poem *To the Many* by Anna Akhmatova, a great Russian poet of the first half of the twentieth century.

21. Norma Farber, *A Desperate Thing*, The Plowshare Press Incorporated, 1973, p. 21.

22. Einstein's letter to Phyllis Wright, January 24, 1936, as quoted in Walter Isaacson, *Einstein: His Life and Universe*, Simon & Schuster, 2007, p. 388.

23. David Brewster, *Memoirs of the Life, Writings, and Discoveries of Sir Isaac Newton*, vol. 2, Adamant Media Corporation, 2001 (reprint of a 1855 edition by Thomas Constable and Co.), p. 407.

Epilogue

1. Edward Frenkel, Robert Langlands, and Ngô Bao Châu, *Formule des Traces et Foncto-rialité: le Début d'un Programme*, Annales des Sciences Mathématiques du Québec **34** (2010) 199–243, available online at http://arxiv.org/pdf/1003.4578.pdf

Edward Frenkel, *Langlands Program, trace formulas, and their geometrization*, Bulletin of AMS, vol. 50 (2013) 1–55, available online at http://arxiv.org/pdf/1202.2110.pdf

Glossary of Terms

Abelian group. A group in which the result of multiplication of any two elements does not depend on the order in which they are multiplied. For example, the circle group.

Automorphic function. A particular kind of function that appears in harmonic analysis.

Automorphic sheaf. A sheaf that replaces an automorphic function in the geometric Langlands relation in the right column of Weil's Rosetta stone.

Category. An algebraic structure comprised of "objects" and "morphisms" between any pair of objects. For instance, vector spaces form a category, and so do sheaves on a manifold.

Circle. A manifold that may be described as the set of all points on the plane equidistant from a given point.

Circle group. The group of rotations of any round object, such as a round table. It is a circle with a special point, the identity element of this group. The circle group is the simplest example of a Lie group.

Complex number. A number of the form $a + b\sqrt{-1}$, where a and b are two real numbers.

Composition (of two symmetries). The symmetry of a given object obtained by applying two symmetries of that object one after another.

Correspondence. A relation between objects of two different kinds, or a rule that assigns objects of one kind to objects of another kind. For example, a one-to-one correspondence.

Cubic equation. Equation of the form $P(y) = Q(x)$, where $P(y)$ is a polynomial of degree two and $Q(x)$ is a polynomial of degree three. An example, which is studied in detail in this book, is the equation $y^2 + y = x^3 - x^2$.

Curve over a finite field. An algebraic object comprised of all solutions of an algebraic equation in two variables (such as a cubic equation) with values in a finite field of p elements and all of its extensions.

Dimension. The number of coordinates needed to describe points of a given object. For example, a line and a circle have dimension one, and a plane and a sphere have dimension two.

Duality. Equivalence between two models (or theories) under a prescribed exchange of parameters and objects.

Fermat's Last Theorem. The statement that for any natural number n greater than 2, there are no natural numbers x, y, z such that $x^n + y^n = z^n$.

Fibration. Suppose we have two manifolds M and B, and a map from M to B. For any point in B, we have the set of points in M mapping to this point, called the "fiber" over this point. M is called a fibration (or a fiber bundle) over the "base" B if all of these fibers can be identified with each other (and each point in B has a neighborhood U whose preimage in M can be identified with the product of U and a fiber).

Finite field. The set of natural numbers between 0 and $p - 1$, where p is a prime number, or its extension obtained by adjoining solutions of a polynomial equation in one variable.

Function. A rule that assigns a number to each point of a given set or manifold.

Fundamental group. The group of all continuous closed paths on a given manifold starting and ending at a given point.

Galois group. The group of symmetries of a number field preserving the operations of addition and multiplication.

Gauge group. A Lie group that appears in a given gauge theory and determines, in particular, the particles and the interactions between them within that theory.

Gauge theory. A physical model of a particular kind, describing certain fields and interactions between them. There is such a theory (or model) for any Lie group, called a gauge group. For example, the gauge theory corresponding to the circle group is the theory of electromagnetism.

Group. A set with an operation (variably called composition, addition, or multiplication) that assigns an element of this set to any pair of elements. (For example, the set of all integers with the operation of addition.) This operation must satisfy the following properties: the existence of the identity element, the existence of the inverse of each element, and associativity.

Harmonic analysis. A branch of mathematics studying decomposition of functions in terms of harmonics, such as the sine and cosine functions.

Hitchin moduli space. The space (or manifold) whose points are representations of the fundamental group of a given Riemann surface in a given Lie group.

Integer. A number that is either a natural number, or 0, or the negative of a natural number.

Kac–Moody algebra. The Lie algebra of the loop group of a given Lie group, extended by an extra line.

Langlands dual group. A Lie group assigned to any given Lie group G by a special procedure. It is denoted by ^LG.

Langlands relation (or Langlands correspondence). A rule assigning an automorphic function (or an automorphic representation) to a representation of a Galois group.

Lie algebra. The tangent space to a Lie group at the point corresponding to the identity element of this group.

Lie group. A group that is also a manifold, such that the operation in the group gives rise to a smooth map.

Loop. A closed curve, such as a circle.

Manifold. A smooth geometric shape such as a circle, a sphere, or the surface of a donut.

Map from one set (or manifold), M, to another set (or manifold), N. A rule that assigns a point of N to each point of M. (A map is sometimes referred to as a mapping.)

Modular form. A function on the unit disc satisfying special transformation properties under a subgroup of the group of symmetries of the disc (called the modular group).

Natural number. Number 1 or any number obtained by adding 1 to itself several times.

Non-abelian group. A group in which the result of multiplication of two elements depends in general on the order in which they are multiplied. For example, the group $SO(3)$.

Number field. A numerical system obtained by adjoining to rational numbers all solutions of a finite collection of polynomials in one variable whose coefficients are rational numbers.

Polynomial in one variable. An expression of the form $a_n x^n + a_{n-1} x^{n-1} + \ldots + a_1 x + a_0$, where x is a variable and $a_n, a_{n-1}, \ldots, a_1, a_0$ are numbers. Polynomials in several variables are defined similarly.

Polynomial equation. An equation of the form $P = 0$, where P is a polynomial in one or more variables.

Prime number. A natural number that is not divisible by any natural number other than itself and 1.

Representation of a group. A rule that assigns a symmetry of a vector space to each element of a given group so that some natural properties are satisfied. More generally, a representation of a group G in another group H is a rule that assigns an element of H to each element of G, so that some natural properties are satisfied.

Quantum field theory. This term may refer to one of two things. First, it may be a branch of physics that studies models of interactions of quantum particles and fields. Second, it may be a *particular* model of this type.

Set. A collection of objects, such as the set $\{0, 1, 2, \ldots, N - 1\}$ for a given natural number N.

Sheaf. A rule that assigns a vector space to each point of a given manifold, satisfying certain natural properties.

Shimura–Taniyama–Weil conjecture. The statement that there is a one-to-one correspondence between cubic equations and modular forms satisfying certain properties. Under this correspondence, the numbers of solutions of the cubic equation modulo prime numbers are equal to the coefficients of the modular form.

SO(3). The group of rotations of a sphere.

Sphere. A manifold that may be described as the set of all points in a flat three-dimensional space that are equidistant from a given point.

Supersymmetry. A type of symmetry in a quantum field theory that exchanges bosons and fermions.

Symmetry. A transformation of a given object that preserves its properties, such as its shape and position.

Theory. A particular branch of mathematics or physics (such as number theory) or a specific model describing relations between objects (such as gauge theory with gauge group *SO*(3)).

Vector space. The set of all vectors in a given n-dimensional flat space, carrying operations of addition of vectors and multiplication of vectors by numbers, satisfying natural properties.

Index